SILVER NANOPARTICLES:
Synthesis, Functionalization and Applications

Authored by

Parteek Prasher

Department of Chemistry, Energy Acres
University of Petroleum & Energy Studies
India

&

Mousmee Sharma

Department of Chemistry, Arcadia Grant
Uttaranchal University
India

SILVER NANOPARTICLES: Synthesis Functionalization and Applications

Authors: Parteek Prasher and Mousmee Sharma

ISBN (Online): 978-981-5050-53-0

ISBN (Print): 978-981-5050-54-7

ISBN (Paperback): 978-981-5050-55-4

need for a court order if at any point you breach any terms of this License Agreement. In no event will any delay or failure by Bentham Science Publishers in enforcing your compliance with this License Agreement constitute a waiver of any of its rights.

3. You acknowledge that you have read this License Agreement, and agree to be bound by its terms and conditions. To the extent that any other terms and conditions presented on any website of Bentham Science Publishers conflict with, or are inconsistent with, the terms and conditions set out in this License Agreement, you acknowledge that the terms and conditions set out in this License Agreement shall prevail.

Bentham Science Publishers Pte. Ltd.
80 Robinson Road #02-00
Singapore 068898
Singapore
Email: subscriptions@benthamscience.net

BENTHAM SCIENCE

CONTENTS

FOREWORD

The book "SILVER NANOPARTICLES: Synthesis, functionalization and Applications" highlights the glorious journey traversed by AgNPs in the contemporary era where they have cemented their place as the most extensively exploited material of the present generation. The most popular theranostic application, tumor annihilating properties, and their prevalence as next-generation antibiotics have been presented epigrammatically. The silver nanoparticles serve as ubiquitous material of the present generation with applications in a wide-array of disciplines that include renewable energy, optoelectronics, image contrast agents, tomography, magnetic imaging, bioprobes, diagnostic kits, gene delivery, and molecular medicine. Among their concurrent applications, the biological interventions of AgNPs have revolutionized the pharma sector by making a significant contribution starting from the drug delivery and ending at bioimaging. The surface engineering of AgNPs causes them to develop functional head-groups that serve as precursors for the chemical transformations to therapeutically relevant molecules with value-added physicochemical and physiological properties in the form of optical, electrical, and magnetic characteristics. This book is a rich blend of professional writing and a raw understanding of the key details of AgNPs in terms of their surface engineering and applications. Apparently, being the most extensively exploited material of the generation with ubiquitous applications, the present readership must inculcate the basic understanding of the synthesis and further utilization of AgNPs, the need that this book mainly caters.

Parteek Prasher
Department of Chemistry
University of Petroleum & Energy Studies
Dehradun
India

PREFACE

The book "SILVER NANOPARTICLES: Synthesis, functionalization and Applications" presents a succinct coverage of the current synthesis protocols, functionalization techniques, and state-of-the-art applications of surface functionalized AgNPs in the medical field. The presented book traverses the journey of AgNPs from synthesis paradigm to functionalization, which further extends to the key biological applications. This book focuses on a wider audience, including medicinal chemists, drug design experts, biological and translational researchers, and physical chemists working in the field of biological nanoscience. The book provides a rich experience of understanding the synthesis, functionalization, and applications of AgNPs in a single reading enriched with self-explanatory images, illustrations, and tables wherever necessary. The book provides useful information for the postgraduate (Master's and research higher degree) doctoral students and postdoctoral research fellows working on the development of advanced nanostructures and nanopharmaceuticals. This book intends to cover both the academic and research requirements of the students for pursuing course work in various academic degree programs, including Physical Sciences, Pharmaceutical Sciences, and various disciplines under the Natural Sciences.

CONSENT FOR PUBLICATION

Not applicable.

CONFLICT OF INTEREST

The author declares no conflict of interest, financial or otherwise.

ACKNOWLEDGEMENTS

Declared none.

Parteek Prasher
Department of Chemistry
University of Petroleum & Energy Studies
Dehradun
India

CHAPTER 1

Silver Nanoparticles: A Ubiquitous Material of Future

Abstract: Silver nanoparticles (AgNPs) offer ubiquitous applications in diverse fields. The materials based on AgNPs provide profitable solutions to the contemporary exigencies such as energy harvesting, future antibiotics, molecular sensors, tracers, image enhancers, tomography contrast agents, antimicrobial fabrics, smart wearables, and many more. The unique physicochemical and optoelectronic properties of AgNPs play a central role in offering these wide applications. The synthesis and stabilization of AgNPs *via* physiologically benevolent molecules, and biomolecules; in addition to the surface engineering of AgNPs to obtain tailored properties, further support the diverse applications of silver-based nanomaterials. The green synthesis of AgNPs by plant extracts, microbial exopolysaccharides, and supernatant provides an economical approach for large-scale generation of AgNPs. The silver nanotechnology popularized after its utilization in household appliances, including water purifiers, washing appliances. However, excessive usage with limited regulations led to the manifestation of increased toxicity in the ecosystem, and successive accumulation in autotrophs and higher trophic levels in a food chain. Therefore, the ubiquitous applications of AgNPs must judiciously consider the environmental impact and influence on the ecosystem to ensure sustainable development. This chapter presents important highlights on the various applications of AgNPs in the modern era.

Keywords: AgNPs, Biosensors, Oligodynamic effect, Opto-electronic properties, Photocatalysis, Solar cells, Surface fabrication, Surface plasmon resonance.

1. INTRODUCTION

The term 'nano' refers to materials having at least one dimension in the range 1-100 nanometer. Metallic particles of spherical, cylindrical, triangular, or cubic shape with dimensions in the same range represent metal nanoparticles, which due to their unique physicochemical and optical properties, present numerous applications. Compared to the bulk matter, the nanoparticles demonstrate a high fervor in developing advanced materials. The noble metal nanoparticles possess unique surface plasmon resonance (SPR) that affords characteristic optical properties useful in biological probing and for developing lab-on-a-chip sensors. The nano-size of the metallic particles enables their entry to the microbial cells

Parteek Prasher & Mousmee Sharma

and tissues where the triggering of redox imbalance by the nanoparticles results in the microbicidal action. In addition to oxidative stress, the metallic nanoparticles instigate a cascade of processes *in vivo* thereby displaying the oligodynamic effect, useful for the antimicrobial activity. Among most of the noble metals, the silver nanoparticles (AgNPs) present ubiquitous applications in diverse fields making them as the most exploitable material of the contemporary era. Starting from optoelectronics to smart devices, from cosmetics to textiles, from fertilizers to pesticides, from solar cells to energy efficient materials, and from next generation antibiotics to drug delivery vectors, the AgNPs dominate over other metallic nanoparticles. However, most of the nanoformulations of AgNPs suffer limitations such as oxidation to Ag^+ ions that exhibit toxicity to the living organisms and their complex physicochemical transformations in environment adversely effects the agroecosystems. Though surface functionalization and fabrication mitigate the toxicity of AgNPs to some extent, however the subsequent loss of surface capping results in direct exposure of AgNPs thereby producing inevitable effects. In this chapter, we will discuss about the ubiquitous applications of AgNPs *via* a *via* toxicity and disposal limitations.

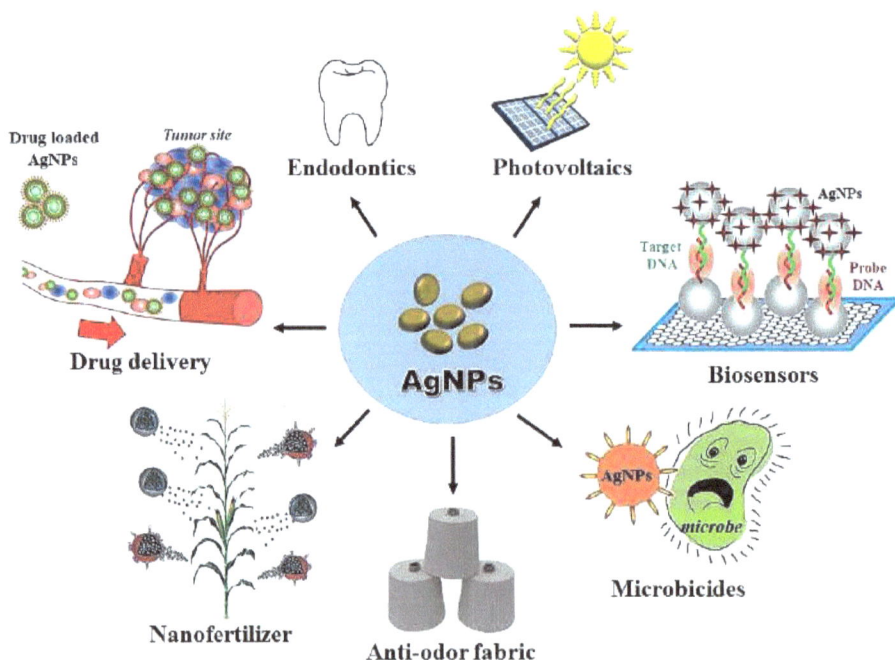

Fig. (1). Contemporary applications of silver nanoparticles (AgNPs).

The shape, size, and surface coating or functionality decide the physicochemical characteristics of AgNPs, which further cater to multiple applications in catalysis, sensing, theranostics, biocides, and energy harvesting. The diverse shapes and size of nanoparticles arise from the synthesis methodology, including several critical physical parameters. Mainly, the presence of a stabilizing agent that preferentially binds to a particular crystal surface arrests the isotropic growth thereby resulting in unidirectional crystal growth and anisotropic shapes. Likewise, the spherical nanoparticles adopt a variety of diameters in the presence of same reducing and capping agent, depending on the temperature, pH of the solution and nature of the solvent. Similarly, the electrical, optical, magnetic, and mechanical properties of the nanoparticles differ markedly as compared to the bulk matter.

1.1. Electrical Properties of AgNPs

The electric properties of AgNPs differ from that of the bulk material due to the restricted movement of electrons at the nanoscale that results in the changes in electric properties of nanomaterial. As such, the semiconductor materials behave as conductors at nanoscale. Similarly, an increase in the concentration of AgNPs improves their electrical conductivity, however the AgNPs of size <10 nm do not conduct electricity, compared to that of the silver metal. Importantly, the conversion of AgNPs to silver nanowire markedly improved the electrical properties and power transfer. The incorporation of AgNPs to polymers such as polyaniline improves the electrical conductivity of the latter due to enhanced electrical mobility due to the presence of AgNPs. Decoration of graphene nanosheets with AgNPs improves the electrical properties of the former by acting as the nanospacer that increases the interlayer distance and improves the electrical conductance between the layers [1]. The decoration of polyester fabric with AgNPs dramatically improves its electrical conductivity due to the 'silver colloid effect". The enhanced conductivity fabric experienced six-fold increase in the conductivity in the presence of AgNPs due to the rearrangement and quantum tunneling effect of AgNPs at nanoscale [2]. Similarly, the addition of AgNPs to electronically conducting adhesives prevents their agglomeration thereby improving the electrical conductivity of the latter [3]. Enhanced electrical performance of low temperature screen-printed AgNPs presented high frequency electronic applications due to a superior packing of AgNPs, thereby reducing the surface roughness by three-fold [4]. The decoration of carbon nanotubes with AgNPs filler significantly improved the electrical conductivity of the CNT-polymer composites by four-fold compared to the pristine and functionalized carbon nanotubes. These findings confirm the unique electrical conductivity of AgNPs for diverse applications [5]. Singh *et al.* (2014) reported the zeta potential and electrical properties of multishaped AgNPs obtained by using maltose as reducing agent in the presence of microwaves. The spherical and anisotropic AgNPs displayed electrical conductivity in the range 2.04×10^{-4} and 1.49×10^{-3} respectively, with anisotropic AgNPs exhibiting superior enhancement in current compared to the spherical AgNPs mainly due to their sharp vertices that promote enhancement in the electrical field [6]. Bhagat *et al.* (2015) reported electrical properties of green synthesized AgNPs dispersed in distilled water and DPPH solution. Notably, the AgNPs dispersed in DPPH solution displayed a rapid increase in the current with respect to the applied voltage, compared to the dispersion of AgNPs in distilled water. The current-voltage characteristics suggested that the changes in current depends on the concentration of DPPH and the current increases with the increase in the antioxidant characteristics of AgNPs. AgNPs@polydopamine core-shell nanoparticles presented applications as fillers into poly(vinylidene difluoride) matrix to prepare dielectric composites, where the core-shell nanoparticles improved the dielectric constant of composites. The $AgNO_3$/dopamine ratio and pH of the dopamine solution displayed significant effect on dielectric properties of composites, with highest dielectric constant achieved at 25 wt percent filled loading at 100 Hz. Besides, the AgNPs display significant thermoelectric properties [7]. Wang *et al.* (2014) reported the thermoelectric properties of AgNPs-polyaniline hybrid nanocomposites, where the electrical performance of the nanocomposite improved markedly on addition of AgNPs. The presence of AgNPs lowered seebeck coefficient, whereas the electric conductivity improved significantly. Negligible enhancement appeared for the thermal conductivity even on increasing the content of $AgNO_3$, which resulted in an improvement of the figure of merit of the nanocomposites, with a maximum value 5.73×10^{-5}, compared to that for the pure polyaniline nanocomposites. These investigations validated the hybridization of conducting polymers with AgNPs as an effective strategy for obtaining improved thermoelectric properties [8]. Shivananda *et al.* (2020) reported the electrical properties of composite films obtained by hybridization of AgNPs with silk-fibroin. The AgNPs reportedly improve the dielectric and AC conductivity of the silk-fibroin film and extend their biosensing applications as implantable thermoelectric wireless switching devices. The AgNPs adopt a spherical shape and crystalline FCC structure with a uniform distribution in the silk-fibroin matrix. The electrical properties of composite film further improved on increasing the concentration of AgNPs, without altering the thermal and mechanical properties of the former [9]. Zhang *et al.* (2016) reported the dielectric properties of polymer composites loaded with polydopamine@AgNPs core-satellite particles, where the presence of AgNPs and the size of polydopamine core markedly enhanced the dielectric constant of the former. The AgNPs reportedly scattered on the polydopamine surface while forming the core-satellite structure, which prevents aggregation of AgNPs and provides stability to the structure [10].

Song *et al.* (2016) reported the effect of *in situ* formed AgNPs on electrical properties of silver nanowires loaded epoxy resin. The AgNPs imparted isotropic electrical properties in epoxy resins containing AgNWs as fillers, whereas the absence of AgNPs demonstrated anisotropic electrical properties to the nanocomposite. Similarly, the presence of AgNPs in combination with boron nitride nanosheets synergistically influences the electrical conductance of epoxy nanocomposites. The epoxy nanocomposites loaded with AgNPs and boron nitride nanosheets display an excellent thermal conductivity, superior insulation strength, low dielectric loss and lesser permittivity [11]. As compared to the pure epoxy, the nanocomposites containing binary nanofillers afford a commendable enhancement in breakdown voltage mainly due to the formation of AgNPs conducting channels that significantly increase the breakdown path. The nanofillers with appreciable electrical conductivity do not affect the insulation properties of nanocomposites upon blocking of the former with electrically insulating nanofillers [12]. The doping of carbon nanofibers with AgNPs reportedly improves the electrical conductivity of the former, which depends on the average diameter of nanofibers and the concentration of doped AgNPs. The total average current improved by five-fold with the increase in the concentration of AgNPs from 30-40 mmol. The increase in electrical conductivity mainly appeared due to the lowering of electrical resistance of carbon nanofibers on adding AgNPs [13]. Gnidakouong *et al.* (2019) appraised the influence of low-temperature sintering on the electrical performance of AgNP-CNT nanocomposites. Reportedly, at a given temperature, the melted aggregates of AgNPs facilitate the joining of CNT ropes, thereby improving their surface electrical properties and reducing the interfacial resistance and tunneling resistance due to π-π electronic interactions. Similarly, the point-like AgNP fillers and advanced sintering resulted in lowering of surface electrical resistance mainly due to the reduction in porosity caused by the coalescence of AgNPs [14]. Bhadra *et al.* (2019) reported the influence of humidity on the electrical properties of AgNP based nanocomposites. The electrical properties of silver-polyaniline/ polyvinyl alcohol nanocomposites displayed uniform changes in resistivity with an increase in humidity. Mechanistically, the water molecules present in humidity donate electrons to the valence bond of polyaniline/ polyvinyl alcohol molecules, thereby increasing the bandgap by lowering the number of holes. In addition, the hydrophobic nature of polyvinyl alcohol causes a swelling effect, which increases the inter-particle distance between the conducting fillers, thereby reducing the overall conductivity of nanocomposites [15].

1.2. Optical Properties of AgNPs

AgNPs absorb and scatter light uniquely depending on the shape, size, and refractive index of the nanoparticles. The AgNPs display surface plasmon resonance (SPR) occurring due to the electromagnetic field induced oscillation of the electrons present in the conduction band. The unique optical properties offer optical biosensing applications to AgNPs to monitor the complex biological processes and for investigating biomolecular interactions. The biosensors based on SPR display extraordinary sensitivity and responsiveness to minute changes occurring in the refractive index of the analyte. Such biosensors provide applications for the detection of biomacromolecules including nucleotides, peptides, antibodies, and enzymes. The optical properties of the AgNPs and their unique localized-SPR bands enable the prediction of nanoparticle size [16]. Similarly, the surface coating of AgNPs fine-tunes their optical properties for enhanced biological applications with reduced toxicity [17]. In addition to the biological applications, the tunable optical properties of hydrophilic AgNPs enable the detection of heavy metals in water. The wavelength of SPR peak for AgNPs displayed a weak dependence on the heavy metal ion concentration in water, where the broadening of SPR band and high background absorption restricted the detection of ions. Tuning of the AgNPs surface overcomes these limitations. Mainly, the hydrophilic AgNPs displayed enhanced sensitivity and selectivity towards Ni^{+2} ions, with a sensitivity of 0.3 ppm [18]. Sivanesan *et al.* (2011) reported citrate functionalized AgNPs with tunable SPR properties for applications in protein analysis and the detection of specific protein cofactors such as cytochrome c, in nanomolar concentration with the help of surface enhanced resonance Raman (SERR) [19]. Raj *et al.* (2017) developed a localized SPR-based dopamine sensor based on L-tyrosine-capped AgNPs. The nanoparticles demonstrated a lowering in fluorescence intensity and an increase in the absorption spectra with an increase in the concentration of dopamine from 0-50 µM. The sensor exhibited a superior sensitivity and selectivity towards dopamine, compared to the other biomolecules with a detection limit of 0.16 µM [20, 21]. Ajitha *et al.* (2016) reported the role of capping agent in the optical properties of AgNPs for the detection of hydrogen peroxide. The loading of polyvinyl alcohol functionality on AgNPs provided localized-SPR based sensor for the detection of H_2O_2 with a detection limit of 10^{-7} M. The strength of localized-SPR changed with changes in the concentration of H_2O_2 and reaction time due to the catalytic degradation of AgNPs [22]. Li *et al.* (2017) investigated the optical limiting properties of AgNPs hybridized with polydimethylsiloxane (PDMS). The optimal limiting effect mainly arises due to non-linear optical adsorption and refraction. Notably, the AgNPs-PDMS hybrid sheets demonstrated superior optical limiting properties compared to the $Ag@SiO_2$ solution mainly due to the early onset of limiting and a wide reduction in transmittance [23]. Adamiv *et al.* (2014) reported the non-linear optical properties of AgNPs. The AgNPs in size range 14-18 nm upon annealing with $Li_2B_4O_7$:Ag glass locate themselves in the thin near-surface layer, thereby forming the interface region, which transforms the positive character of non-linear refraction of $Li_2B_4O_7$:Ag glass to negative. The change markedly enhances its non-linear properties due to plasmon resonance [24]. Nisha *et al.* (2019) further reported the optical limiting behavior of AgNPs. The nanoparticles displayed a non-linear refractive index of 7.15 x 10^{-8} cm^2/W and the non-linear absorption coefficient of 0.04 x 10^{-4} cm^2/W, whereas the third order non-linear susceptibility appeared as 4.30 x 10^{-6} esu [25]. Pandey *et al.* (2012) reported biopolymer-AgNPs nanocomposites for the optical detection of ammonia. The SPR-based sensor displayed a response time of 2s, and exhibited a detection limit of 1ppm towards ammonia solution at room temperature. The localized-SPR properties of the dispersion of AgNPs in the polymer matrix showed calorimetric sensing applications with a high reproducibility and fast response time. The optical sensor based on AgNPs displayed application for the physiological detection of ammonia in biological fluids, including plasma, cerebrospinal fluid, saliva, and sweat [26]. Edison *et al.* (2016) developed an AgNPs based optical sensor for the detection of dissolved ammonia. The appearance of SPR peak and yellowish color revealed the formation of AgNPs in control solution and ammonia containing solutions. Further, the SPR absorbance of AgNPs in ammonia containing solutions increased with the formation of diamine silver complex and in the presence of ammonium phenolate ions. These events increased the rate of AgNPs nucleation and yielded small-sized nanoparticles of average diameter 30 nm with a distorted spherical shape, in the absence of ammonia. The presence of 100-ppm ammonia solution, however yielded spherical shaped AgNPs with an average diameter of 5 nm [27]. Bhutto *et al.* (2018) reported the plasmonic properties of AgNPs obtained from the phenolic compounds as reducing agents. Reportedly, the AgNPs display higher plasmonic response, which depends on the antioxidant properties of phenolic acids and the degree of hydroxylation of the latter. The higher degree of hydroxylation resulted in ameliorated scavenging property of the phenolate compounds and a higher potency to reduce Ag^+ ions to AgNPs. Similarly, the rate of reaction and reducing power depended on the nature of phenolate and its substituent pattern [28]. Singha *et al.* (2014) reported the synthesis of high optical quality AgNPs by the reduction of ascorbic acid at room temperature. The obtained AgNPs displayed extremely sharp and intense SPR bands with narrow bandwidth, superior to the conventional bands obtained from the AgNPs synthesized by common reducing agents [29]. Kemper *et al.* (2017) studied the effect of LED irradiation on the optical properties of AgNPs in the polyethylenimine thin films. The tailoring of AgNPs in the transparent polymer matrix provided novel applications, including the adaptable light filters for the perspective lab-on-a-chip applications. Reportedly, the AgNPs begin reshaping their morphology upon irradiation, thereby resulting in the changes in absorption signals and SPR properties. Green light irradiation leads to the forced plasmon oscillation by the excited regions on AgNPs, resulting in the photoreduction of redundant Ag^+ ions [30]. Wang *et al.* (2018) reported the development of polydopamine capped AgNPs based SPR biosensor for a highly sensitive, regenerative and stable detection of horse IgG with a detection limit of 0.625 µg/mL. The AgNPs acted as signal enhancing labels with a 2/4-fold higher detection limit compared to the gold nanoparticles. The SPR based biosensor displayed a high selectivity towards the horse IgG, with a bonding constant of 2.93 x 10^7 L mol^{-1} to the antibody, as detected by the polydopamine loaded AgNPs based SPR biosensor. The desired performance by polydopamine loaded AgNPs film-sensing platform occurred due to the effective antibody immobilization by polydopamine, whereas the presence of AgNPs improves the sensitivity of the biosensor owing to the electronic coupling between AgNPs, hence amplifying the SPR response [31]. Raj *et al.* (2017) developed a highly sensitive and selective SPR-based fiber optic sensor based on AgNPs for cysteine detection. The sensor displayed a detection limit of 7.7 nM for cysteine among various biomolecules. While increasing the cysteine concentration, the resonance wavelength lowered and shifted slightly towards the lower wavelength. Similarly, the localized-SPR resonance peak intensity lowered on adding cysteine, thereby suggesting the event of transduction pertaining to the interactions of cysteine with capped AgNPs [20, 21]. Mota *et al.* (2020) suggested that the low-intensity polychromatic light irradiation markedly governs the optical properties of AgNPs. Reportedly, the resonance between SPR bands of AgNPs governs the self-limiting growth process of anisotropic AgNPs. Importantly, the light emitting diode irradiation wavelength decides the final morphology of the AgNPs due to the shape-dependence of the plasmonic spectrum of AgNPs. These investigations suggested a photoinduced control of the morphology and the plasmonic properties of the AgNPs in the presence of low-intensity light emitting diode [32]. Karimzadeh *et al.* (2010) investigated the non-linear optical properties of AgNPs in water with a continuous wave laser irradiation at 532 nm. The closed Z-scan measurements suggested the thermal effect on the non-linear refractive index of AgNPs. The aberrant thermal lens model follows in agreement with the Z-scan investigations for AgNPs, with a nonlinear refractive index of -1.0 x 10^{-8} cm^2/W. and thermo-optic coefficient as -0.99 x 10^{-4} W/mK. These investigations revealed that the thermal nonlinear effects play a significant role in deciding the photonic applications of AgNPs and for appraising their nonlocal nonlinear processes [33]. Pugazhendi *et al.* (2015) reported nonlinear optical properties of AgNPs obtained from *Alpinia calcarata*. Nonlinear optical studies performed by single beam Z-scan setup optimized the nonlinear refractive index of the order 10^{-8} cm^2W^{-1}, whereas the nonlinear absorption coefficient and the third order nonlinear susceptibility appeared as 10^{-3} cmW^{-1} and 10^{-3} esu. These investigations suggested that AgNPs demonstrate a superior optical non-linearity as evidenced by the Z-scan technique [34]. Zhang *et al.* (2019) investigated the regulation of optical properties of triphenylamine-capped AgNPs based on the SPR effect. The nanoparticles display a red shift in the UV-Vis absorption upon interfacial coordination caused due to an increase in the electron withdrawing strength presented by Ag atom. The SPR effect of AgNPs of size 6nm improves the single photon fluorescence emission and two-photon absorption. The triphenylamine-AgNPs hybrids display a higher cross-section for two-photon resonance hence presenting excellent applications in optical power limiting with a threshold value of 0.49 J/cm^{-2}. The reported interfacial coordination induced hybrids afford a favorable approach for effectively regulating the linear optical properties and for optimizing the nonlinear performance [35].

1.3. Catalytic Properties

The AgNPs present excellent photocatalytic properties, which present several applications, including the degradation of organic pollutants and dyes. Jiang *et al.* (2005) reported the catalytic applications of AgNPs supported on silica nanospheres. The supporting of nanoparticles on silica spheres avoids their flocculation during the catalytic activity in the solution, thereby enabling an optimal catalytic activity. Notably, the presence of surfactants lowered the catalytic activity of AgNPs by inhibiting the adsorption of reactant molecules on the surface of nanoparticles. Similarly, the presence of electrolytes in the solution possesses the ability to enhance the rate of migration of the reactants in the solution resulting in the enhancement of the rate of catalytic reaction. However, the electrolyte may also restrict the adsorption of reactant molecules on the surface of AgNPs hence lowering the catalytic activity. Anchoring of AgNPs on silica particles prevents their aggregation and avoids the deactivation or poisoning of catalyst during the catalytic reaction [36]. Chandraker *et al.* (2019) presented photocatalytic properties of biogenic AgNPs obtained from *Ageratum conyzoides*. 2h exposure of AgNPs to methylene blue resulted in the degradation of the dye in the presence of sunlight. Visual detection of color change from blue to colorless confirmed the dye degradation. Mainly, photons contained in the incident solar radiation excited the electrons present on the surface of AgNPs, which further are accepted by the dissolved molecules of oxygen present in the solution. The free electrons convert dioxygen to oxygen anion radicals that eventually catalyze the conversion of dye into simpler organic molecules, hence causing its degradation. AgNPs reportedly displayed photocatalytic oxidation of nitric oxide over the nanoparticle-loaded carbon fiber cloths [37]. The modification of TiO$_2$ by AgNPs stabilized the photoefficiency of the composites during the five consecutive cycles of nitric oxide photooxidation, thereby hindering the formation of NO$_2$. The carbon fiber cloths containing 3.7 wt% of AgNPs displayed the highest removal rate for nitric oxide. The maximum and minimum removal rates of nitric oxide appeared to be 80% and 95%, respectively [38]. Zhang *et al.* (2011) reported the catalytic reduction of 4-nitrophenol by AgNPs. The nanocomposites containing carbon-nanofibers well-dispersed AgNPs demonstrated a high catalytic efficiency mainly due to the high surface area of nanoparticles and synergistic effect of AgNPs on the transfer of electrons between the carbon nanofibers and AgNPs. The catalytic efficiency of nanocomposite further enhanced on increasing the content of AgNPs. This nanocatalyst exhibited high recyclability due to one-dimensional nanostructural morphology [39]. Khan *et al.* (2016) reported improved photocatalytic applications of AgNPs synthesized by green methods. The AgNPs mediated the catalytic photodegradation of bromo phenyl blue dye by a significant 98%. The following application extends towards water purification by converting the hazardous organic dyes to non-hazardous materials. In addition, the AgNPs exhibited electro-catalytic properties by reducing the phenolic compounds to hydroquinone. Importantly, the physicochemical properties such as size, shape and catalytic amount decide the photodegradation of bromo phenyl blue dye [40]. Jishma *et al.* (2016) reported the photocatalytic degradation of Victoria Blue B dye by the biogenic AgNPs. The degradation of the dye depended on initial dye concentration, the concentration of AgNPs and pH of the solution, with approximately 78% of the dye degrading within the first 8h. Notably, the photodegradation by AgNPs is dramatically enhanced in the presence of sunlight [41]. Elemike *et al.* (2016) reported the photocatalytic activity of AgNPs synthesized from *Verbascum Thapsus* for the photodegradation of nitrobenzene. The percentage degradation of nitrobenzene by AgNPs occurred by 79.5-87.5% at 20 and 48 h, respectively, with photocatalytic activity depending on morphology, size of nanoparticles and their crystal structure. The high light absorption potency of AgNPs afford good photodegradation properties due to their morphology and distinctive particle size resulting in high surface area and more active sites. Similarly, the lack of aggregation and high dispersity contributed to the photodegradation effect [42]. Khan *et al.* (2016) reported the ultra-efficient photodegradation of methylene blue dye by 99.24%, *via* biogenic AgNPs. The highly efficient catalytic activity of AgNPs arises from electron relay effect, where the AgNPs act as redox catalyst and regulate the electron transfer from the donor polyphenol molecules to the reactant methylene blue dye. The catalytic potential of AgNPs depended on the reduction potential in the catalytic process. The catalytically active AgNPs possess negative redox potential compared to their bulk forms, which assist in photocatalysis [43]. Zou *et al.* (2012) reported the photocatalytic decomposition of methylene blue by halloysite nanotubes supported AgNPs. The degradation rate of methylene blue by AgNPs appeared to be 10% at 60 minutes. The loading of AgNPs on halloysite nanotubes significantly improved the decomposition of methylene blue dye. The decomposition occurred in two stages and ended with the evolution of CO_2, which increases with the time of irradiation. The addition of polyphenols cause chelation with Ag^+ ions and oxidize further to form quinones, thereby forming AgNPs. In addition, the polyphenol molecules contribute significantly towards the stabilization of AgNPs to form a stable Ag/polymer complex [44]. Li *et al.* (2014) reported a non-aggregation colorimetric assay for appraising the catalytic properties of aptamer functionalized AgNPs. The catalytically active nanoparticles mediated the reduction of rhodamine B dye for thrombin detection. Similarly, the immobilized AgNPs enabled the detection of thrombin through a decrease in absorbance due to rhodamine B. The nanosystem enabled highly sensitive detection of thrombin in the picomolar range with high selectivity over other proteins. The sandwich binding potency of aptamers enabled a superior specificity, whereas the magnetic nanoparticles afforded a rapid capture and separation for the visual detection of thrombin. Moreover, the nanosystem presented applications for molecular and biological targets containing more than one binding site [45]. Priebe *et al.* (2014) reported the catalytic properties of AgNPs encapsulated in silica nanocontainers. Silica layer completely covers AgNPs, which are large enough not to leak out of the silica layer and are homogenously loaded inside silica hollow sphere. The utility of Igepal CO-520 surfactant resulted in the formation of homogeneous AgNPs covered uniformly by a silica shell. Similarly, increasing the concentration of $AgNO_3$ in solution resulted in the formation of large sized AgNPs with diverse shapes. The utility of hydrazine to produce AgNPs resulted in the protection of the latter from the etching properties of aqueous ammonia. The mentioned $Ag@SiO_2$ nanocontainers exhibited excellent catalytic potency to reduce methylene blue dye [46]. Yan *et al.* (2018) reported the catalytic application of AgNPs supported on 2D silica nanosheets for the reduction of 4-nitrophenol. The reduction of 4-nitrophenol with AgNP-silica nanosheets nanocatalyst was completed in 40s without mechanical agitation at a rate constant of 80.19×10^{-3} s^{-1}. The nanocatalyst displayed a high turnover frequency of 3.52 min^{-1} hence suggesting a high-density dispersion of small sized and uniformly distributed AgNPs. In addition, the nanocatalyst offered multiple active sites for making optimal contact with reactants and a swift transfer of the interfacial electrons from the surface of AgNPs to the surface of 4-nitrophenol. The 2D silica nanosheets presented potential application as support material for a highly efficient, uniform assembly of AgNPs. The 2D nanocomposites exhibited stable catalytic efficiency of nearly 100% over five cycles of catalytic reaction. The nanocatalyst held a high efficiency for the degradation of 4-nitrophenol, which is further purified from the contaminated water sample with the help of convenient filtering and catalyzing device which detects the degradation of 4-nitrophenol colorimetrically [47].

1.4. Photovoltaic Properties

The AgNPs exhibit excellent photovoltaic properties by solar light harvesting and converting it into useful energy. Liu *et al.* (2013) reported improved photovoltaic performance of organic hybrid solar cell in the presence of AgNPs due to their unique light scattering properties. The AgNP-decorated SiNW/organic hybrid solar cells, treated as double junction tandem solar cells displayed an enhanced performance due to current enhancement by AgNPs. The nanoparticles increased the short circuit current from 10.5 mA/cm^2 to 16.6 mA/cm^2 showing an enhancement of 58%. Gain in current resulted in the increase of conversion efficiency from 2.47% to 3.23% that occurred mainly due to the surface plasmon scattering properties of AgNPs [48]. Ramadan *et al.* (2020) developed hybrid silver nanostructures for enhanced photovoltaic performance. The presence of AgNPs resulted in the improvement of spectral photocurrent response, short circuit current density, open-circuit voltage, fill factor, and photovoltaic efficiency. Reportedly, the ultrafine size of AgNPs in the range 5-15 nm manifested SPR effects in the blue region of electromagnetic spectra. In addition, the AgNPs improved the electrical conduction by nanoporous silicon layers, which produced the desired photovoltaic effect [49]. Joshi *et al.* (2018) reported multi-shaped AgNPs for efficient light harvesting properties in dye-sensitized solar cells. The light harvesting properties arise due to tunable and broadband surface plasmon-resonance properties. The presence of AgNPs improved the photo-conversion efficiency of N719 dye-sensitized solar cells by 30%. Further, the presence of AgNPs in thin films improved the charge separation process, which resulted in enhanced electron transport, thereby increasing the overall efficiency in plasmonic-based dye-sensitized solar cells [50]. Singh *et al.* (2018) reported solution processed AgNWs based transparent conductive electrode for highly efficient photovoltaic performance. The AgNWs provided a suitable platform for solar cell fabrication. The photovoltaic efficiency suffered limitations owing to the poor adhesion and small contact area of AgNWs, the further loading of ZnO films generated AgNWs-based transparent conducting electrode with mass density 242 mg/m^2, sheet resistance of 11 Ω/sq with 90% transmittance. In addition, the AgNWs-based transparent conducting electrode displayed a maximum conversion efficiency of 13.4%, which is highly favorable compared to the customary Indium tin oxide based TCEs [51]. Liu *et al.* (2014) reported the role of AgNPs in generating the plasmonic-enhanced, highly efficient polymeric solar cells. The blending of differently sized AgNPs into the anode buffer layer triggered localized-SPR resulting in an enhanced broadband absorption of polymeric solar cells. The double anode buffer layers reduced the surface roughness of composite buffer layer, which played a vital role in demonstrating the electrical properties of polymeric solar cells. The AgNPs impregnated polymeric solar cells displayed a power conversion efficiency of 9.2% compared to the pre-optimized control PSCs. These findings suggested a novel approach to achieve a higher overall efficiency pertaining to the cooperative plasmonic effect arising from the dual resonance enhancement caused by differently sized AgNPs [52]. Joshi *et al.* (2017) extended the same strategy towards dye-sensitized solar cells whose light harvesting efficiency enhanced in the presence of differently sized AgNPs, mainly due to broadband surface plasmon resonance. The presence of 1-wt % of AgNPs in 4μm thin reference photoanode improved the photon conversion efficiency of dye-sensitized solar cells by 50%. The improvement mainly arises due to an amelioration in light harvesting potency of plasmonic photoanode caused by enhancement in local electromagnetic fields, in addition to the stronger near filed scattering effect of AgNPs. Notably, the reduction in charge-transfer resistance value further confirmed the generation of high charge carriers in plasmonic dye-sensitized AgNPs [53]. Ghadari *et al.* (2020) reported phthalocyanine-AgNPs structures for achieving plasmon-enhancement in dye-sensitized solar cells. Direct conjugation of AgNPs to phthalocyanine sensitized the solar cells by increasing short-circuit current density and open-circuit voltage mainly due to increased absorption of the dye by plasmonic effect and electron storage in plasmonic AgNPs. The dyes containing carboxyl- and sulfonyl-conjugated groups displayed better performance mainly due to chemisorption and superior electronic interactions with AgNPs. The anchoring of AgNPs in dye-sensitized solar cells improved the photocurrent and photovoltage owing to the light-harvesting and electron storage properties of plasmonic AgNPs [54]. Seoudi *et al.* (2018) further reported effect of the size of AgNPs on photovoltaic performance of phthalocyanine-sensitized thin films. The changes in particle size result in alteration of SPR bands. The photovoltaic efficiency of these thin films appeared in the range 0.237% to 0.280% for AgNPs with particle size in the range 6nm to 14nm. The efficiency increased with increase in particle size of AgNPs mainly due to enhanced scattering and low reflection of the incident radiation at the thin film surface [55]. Zhang *et al.* (2014) developed silver nanoprisms and appraised their effect in enhancing the photovoltaic performance of organic solar cells. The incorporation of organic solar cells with AgNPs improved the power conversion efficiency from 23.60% to 39.80%. Further, the short-circuit current density of solar cells improved by 17.44% to 10.84 mAcm^{-2} [56]. Su *et al.* (2015) reported graphene nanosheets/ AgNPs nanohybrids for enhanced photovoltaic performance of polymeric solar cells. The nanohybrids increased the power conversion efficiency and photocurrent density of solar cells from 4.04% to 19.5% owing to the localized-SPR of AgNPs, in addition to an improved compatibility and charge transfer capacity between the cathode and photoactive layer [57]. Metzman *et al.* (2019) investigated the role of polystyrene layer on plasmonic silver nanoplates in organic solar cells. Varying concentration of solution containing polystyrene-AgNPs resulted in varying densities on the active layer. The localized-SPR of AgNPs improved the light absorption of active layer eventually leading to an increase in the yield of excitons as indicated by increase in the intensity of photoluminescence emission. The addition of AgNPs further lowered the series resistance and improved the photocurrent of organic photovoltaic. Notably, the power conversion efficiency of organic photovoltaic significantly increased in the presence of AgNPs [58]. Higgins *et al.* (2018) reported enhanced reproducibility of perovskite solar cells by doping fullerene with AgNPs. The addition of small concentration of AgNPs in perovskite solar cell layer ameliorated the reproducibility of solar cell fabrication. The plasmonic-electrical effects presented a greater impact compared to the plasmonic-optical effect including the near field enhancement and scattering, which played a leading role in the performance enhancement [59]. Kazmi *et al.* (2017) investigated AgNPs-mediated efficiency enhancement of dye-sensitized solar cells. The presence of AgNPs improved short circuit current density from 6.95 to 12.58 mAcm^{-2} while the fill factor increased the efficiency from 3.29 to 5.61%. The fabrication of dye-sensitive solar cells with AgNPs yielded maximum efficiency compared to non-fabricated cells. Notably, the optimal electron transport and enhanced dye absorption on AgNPs led to the achievement of maximum conversion efficiency of 7.3% [60]. Sreeja *et al.* (2020) reported efficiency enhancement of natural pigment-sensitized solar cells by SPR properties of AgNPs. The plasmon-enhanced solar cells improved the efficiency of pigment-sensitized solar cells by 20% and increased current generation mainly due to an enhanced absorption by the natural pigment and due to the localized-SPR effect displayed by AgNPs. The anchoring of AgNPs increased electron lifetime and lowered the recombination in photoanodes of pigment-sensitized solar cells, mainly due to the SPR effect of AgNPs that resulted in enhanced generation of excitons and improved the charge transfer in the photoanode. These investigations indicated that the inclusion of AgNPs into photoanode improved the overall performance of natural pigment-sensitized solar cells [61].

1.5. Environmental Impact

The excessive usage of silver-based nanomaterials caused a serious environmental impact pertaining to their direct exposure to the ecosystem [62]. The source of silver nanoparticles in the environment include disposed fabrics and appliances, solar cells, laboratory waste, laundry wash-off, and wastewater treatment plants. From these sources, the AgNPs become a part of the natural biogeochemical cycles, or bioaccumulate in the various realms of an ecosystem, which further biomagnify the environmental occurrence of AgNPs. Autotrophs uptake the AgNPs that become a part of soil or surface run off eventually become a part of the higher food chains and intertwine in the ecological food webs [63]. The various physicochemical transformations of AgNPs including oxidation, dissolution, chloridation, sulfidation, and reduction [64] result in the generation of a large number of species with high persistence in the environment [65]. The hazardous effects of AgNPs on human health occur mainly due to the generation of reactive oxygen species (ROS) when exposed to AgNPs and silver ions. These ROS cause imbalance in the redox homeostasis of the body and trigger a cascade of deleterious effects that prove detrimental to the human health. The ROS potentially damage the cell membranes, DNA double helix, cause mitochondrial dysfunction, and ribosomal impairment. An excessive accumulation of silver in the body owing to a prolonged exposure causes a condition 'Argyria' where the mucous membranes or the skin of the person becomes permanently grey [66]. Fig. (**2**) highlights the potential sources of the environmental exposure of AgNPs.

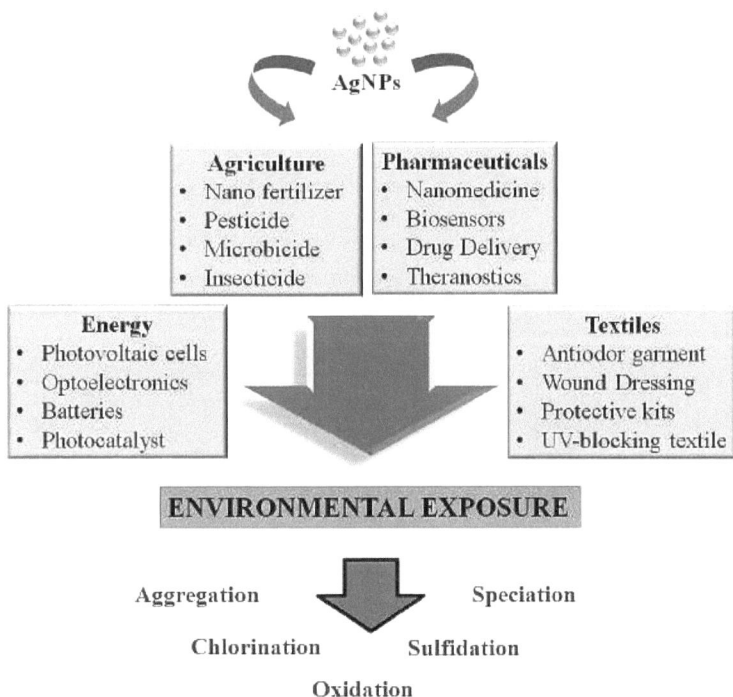

Fig. (2). Sources of the environmental exposure of AgNPs.

CONCLUSION

The ubiquitous applications of AgNPs provided a 'silver lining' in the contemporary era of nanotechnology. The AgNPs find applications in various fields and offer solutions to the most recent exigencies in the present century. The excellent optical, electrical, thermal and oligodynamic properties of AgNPs play a leading role in the determination of their wide applications. A small surface area, engineered surface, fabrication of surface functionalization, synthesis by green methods further support the popularity of AgNPs as the most extensively explored material in the present era. However, the excessive utilization causes serious health and ecological concerns owing to direct exposure of the AgNPs. The disposed materials containing nano silver pose a serious environmental concern as the metallic silver becomes a part of the biogeochemical cycles and accumulate in the higher trophic levels of a food chain while intermingling with the various food webs in an ecosystem. The effective utilization of AgNPs must therefore consider proper disposal and discarding of the nano-silver containing materials to ensure environmental sustainability and non-toxicity to the various components in an ecosystem.

REFERENCES

[1]   Tien, H.W.; Huang, Y.L.; Yang, S-Y.; Wang, J-Y.; Ma, C-C. The production of graphene nanosheets decorated with silver nanoparticles for use in transparent, conductive films. *Carbon,* **2011**, *49*(5), 1550-1560.
        [http://dx.doi.org/10.1016/j.carbon.2010.12.022]

[2]   Zhao, H.; Tian, M.; Li, Z.; Zhang, Y.; Zhu, S.; Zhang, X.; Chen, S.; Qu, L. Enhanced electrical conductivity of silver nanoparticles decorated fabrics with sandwich micro-structure coating layer based on "silver colloid effect". *Mater. Lett.,* **2019**, *240*, 5-8.
        [http://dx.doi.org/10.1016/j.matlet.2018.12.052]

[3]   Chen, D.; Qiao, X.; Qiu, X.; Chen, J. Synthesis and electrical properties of uniform silver nanoparticles for electronic applications. *J. Mater. Sci.,* **2009**, *44*(4), 1076-1081.
        [http://dx.doi.org/10.1007/s10853-008-3204-y]

[4]   Alshehri, A.H.; Jakubowska, M.; Młożniak, A.; Horaczek, M.; Rudka, D.; Free, C.; Carey, J.D. Enhanced electrical conductivity of silver nanoparticles for high frequency electronic applications. *ACS Appl. Mater. Interfaces,* **2012**, *4*(12), 7007-7010.
        [http://dx.doi.org/10.1021/am3022569] [PMID: 23151185]

[5]   Ma, P.C.; Tang, B.Z.; Kim, J-K. Effect of CNT decoration with silver nanoparticles on electrical conductivity of CNT-polymer composites. *Carbon,* **2008**, *46*(11), 1497-1505.
        [http://dx.doi.org/10.1016/j.carbon.2008.06.048]

[6]   Singh, S.; Bharti, A.; Meena, V.K. Structural, thermal, zeta potential and electrical properties of disaccharide reduced silver nanoparticles. *J. Mater. Sci. Mater. Electron.,* **2014**, *25*(9), 3747-3752.
        [http://dx.doi.org/10.1007/s10854-014-2085-x]

[7]   Bhagat, M.; Rajput, S.; Arya, S.; Khan, S.; Lehana, P. Biological and electrical properties of biosynthesized silver nanoparticles. *Bull. Mater. Sci.,* **2015**, *38*(5), 1253-1258.
        [http://dx.doi.org/10.1007/s12034-015-1007-8]

[8]   Wang, W.; Sun, S.; Gu, S.; Shen, H.; Zhang, Q.; Zhu, J.; Wang, L.; Jiang, W. One-pot fabrication and thermoelectric properties of Ag nanoparticles–polyaniline hybrid nanocomposites. *RSC Advances,* **2014**, *4*(51), 26810-26816.
        [http://dx.doi.org/10.1039/C4RA02136C]

[9]   Shivananda, C.S.; Rao, B.L. Sangappa. Structural, thermal and electrical properties of silk fibroin–silver nanoparticles composite films. *J. Mater. Sci. Mater. Electron.,* **2020**, *31*(1), 41-51.

[http://dx.doi.org/10.1007/s10854-019-00786-3]

[10] Zhang, L.; Gao, R.; Hu, P.; Dang, Z-M. Preparation and dielectric properties of polymer composites incorporated with polydopamine@AgNPs core-satellite particles. *RSC Advances,* **2016**, *6*(41), 34529-34533.
[http://dx.doi.org/10.1039/C6RA00827E]

[11] Song, G-S.; Lee, D.S.; Kang, I. The Effects of *in Situ*-Formed Silver Nanoparticles on the Electrical Properties of Epoxy Resin Filled with Silver Nanowires. *Polymers (Basel),* **2016**, *8*(4), 157.
[http://dx.doi.org/10.3390/polym8040157] [PMID: 30979250]

[12] Wu, Y.; Zhang, X.; Negi, A.; He, J.; Hu, G.; Tian, S.; Liu, J. Synergistic effects of Boron nitride (BN) nanosheets and Silver (Ag) nanoparticles on thermal conductivity and electrical properties of epoxy nanocomposites. *Polymers (Basel),* **2020**, *12*(2), 426.
[http://dx.doi.org/10.3390/polym12020426] [PMID: 32059366]

[13] Ali, W.; Shabani, V.; Linke, M.; Sayin, S.; Gebert, B.; Altinpinar, S.; Hildebrandt, M.; Gutmann, J.S.; Mayer-Gall, T. Electrical conductivity of silver nanoparticle doped carbon nanofibres measured by CS-AFM. *RSC Advances,* **2019**, *9*(8), 4553-4562.
[http://dx.doi.org/10.1039/C8RA04594A]

[14] Gnidakouong, J.R.N.; Kim, H. Effects of low-temperature sintering on surface morphology and electrical performance of silver nanoparticle/ carbon nanotube composite film. *Compos., Part B Eng.,* **2019**, *163*, 634-641.
[http://dx.doi.org/10.1016/j.compositesb.2018.12.124]

[15] Bhadra, J.; Popelka, A.; Abdulkareem, A.; Lehocky, M.; Humpolicek, P.; Al-Thani, N. Effect of humidity on the electrical properties of the silver-polyaniline/ polyvinyl alcohol nanocomposites. *Sens. Actuators A Phys.,* **2019**, *288*, 47-54.
[http://dx.doi.org/10.1016/j.sna.2019.01.012]

[16] Amirjani, A.; Firouzi, F.; Haghshenas, D.F. Prediction of the size of silver nanoparticles from their optical properties. *Plasmonics,* **2020**, *15*(4), 1077-1082.
[http://dx.doi.org/10.1007/s11468-020-01121-x]

[17] Fahmy, H.M.; Mosley, A.M.; Elgany, A.A.; Shams-Eldin, E.; Serea, E.S.A.; Ali, S.A.; Shalan, A.E. Coated silver nanoparticles: synthesis, cytotoxicity, and optical properties. *RSC Advances,* **2019**, *9*(35), 20118-20136.
[http://dx.doi.org/10.1039/C9RA02907A]

[18] Prosposito, P.; Mochi, F.; Ciotta, E.; Casalboni, M.; De Matteis, F.; Venditti, I.; Fontana, L.; Testa, G.; Fratoddi, I. Hydrophilic silver nanoparticles with tunable optical properties: application for the detection of heavy metals in water. *Beilstein J. Nanotechnol.,* **2016**, *7*, 1654-1661.
[http://dx.doi.org/10.3762/bjnano.7.157] [PMID: 28144514]

[19] Sivanesan, A.; Ly, H.K.; Kozuch, J.; Sezer, M.; Kuhlmann, U.; Fischer, A.; Weidinger, I.M. Functionalized Ag nanoparticles with tunable optical properties for selective protein analysis. *Chem. Commun. (Camb.),* **2011**, *47*(12), 3553-3555.
[http://dx.doi.org/10.1039/c0cc05058j] [PMID: 21321696]

[20] Raj, D.R.; Sudarsanakumar, C. Surface plasmon resonance based fiber optic sensor for the detection of cysteine using diosmin capped silver nanoparticles. *Sens. Actuators A Phys.,* **2017**, *253*, 41-48.
[http://dx.doi.org/10.1016/j.sna.2016.11.019]

[21] Raj, D.R.; Prasanth, S.; Sudarsanakumar, C. Development of LSPR-Based Optical Fiber Dopamine Sensor Using L-Tyrosine-Capped Silver Nanoparticles and Its Nonlinear Optical Properties. *Plasmonics,* **2017**, *12*(4), 1227-1234.
[http://dx.doi.org/10.1007/s11468-016-0380-5]

[22] Ajitha, B.; Reddy, Y.A.K.; Reddy, P.S.; Jeon, H-J.; Ahn, C.W. Role of capping agents in controlling silver nanoparticles size, antibacterial activity and potential application as optical hydrogen peroxide sensor. *RSC Advances,* **2016**, *6*(42), 36171-36179.

[http://dx.doi.org/10.1039/C6RA03766F]

[23]   Li, C.; Liu, M.; Yan, L.; Liu, N.; Li, D.; Liu, J.; Wang, X. Silver nanoparticles/polydimethylsiloxane hybrid materials and their optical limiting property. *J. Lumin.*, **2017**, *190*, 1-5.
[http://dx.doi.org/10.1016/j.jlumin.2017.05.023]

[24]   Adamiv, V.T.; Bolesta, I.M.; Burak, Y.Y.; Gamernyk, R.V.; Karbovnyk, I.D.; Kolych, I.I.; Kovalchuk, M.G.; Kushnir, O.O.; Periv, M.V.; Teslyuk, I.M. Nonlinear optical properties of silver nanoparticles prepared in Ag doped borate glasses. *Physica B*, **2014**, *449*, 31-35.
[http://dx.doi.org/10.1016/j.physb.2014.05.009]

[25]   Nisha, B.; Vidyalakshmi, Y.; Geetha, D.; Parveen, J.R.; Vinitha, G. Green synthesis, characterization of silver nanoparticles and their study on antibacterial activity and optical limiting behavior. *Appl. Phys. B*, **2019**, *125*(7), 123.
[http://dx.doi.org/10.1007/s00340-019-7226-8]

[26]   Pandey, S.; Goswami, G.K.; Nanda, K.K. Green synthesis of biopolymer-silver nanoparticle nanocomposite: an optical sensor for ammonia detection. *Int. J. Biol. Macromol.*, **2012**, *51*(4), 583-589.
[http://dx.doi.org/10.1016/j.ijbiomac.2012.06.033] [PMID: 22750580]

[27]   Edison, T.N.J.I.; Atchudan, R.; Lee, Y.R. Optical sensor for dissolved ammonia through the green synthesis of silver nanoparticles by fruit extract of *Terminalia chebula*. *J. Cluster Sci.*, **2016**, *27*(2), 683-690.
[http://dx.doi.org/10.1007/s10876-016-0972-4]

[28]   Bhutto, A.A.; Kalay, Ş.; Sherazi, S.T.H.; Culha, M. Quantitative structure-activity relationship between antioxidant capacity of phenolic compounds and the plasmonic properties of silver nanoparticles. *Talanta*, **2018**, *189*, 174-181.
[http://dx.doi.org/10.1016/j.talanta.2018.06.080] [PMID: 30086903]

[29]   Singha, D.; Barman, N.; Sahu, K. A facile synthesis of high optical quality silver nanoparticles by ascorbic acid reduction in reverse micelles at room temperature. *J. Colloid Interface Sci.*, **2014**, *413*, 37-42.
[http://dx.doi.org/10.1016/j.jcis.2013.09.009] [PMID: 24183428]

[30]   Kemper, F.; Beckert, E.; Ebenhardt, R.; Tunnermann, A. Light filter tailoring – the impact of light emitting diode irradiation on the morphology and optical properties of silver nanoparticles within polyethylenimine thin films. *RSC Advances*, **2017**, *7*(66), 41603-41609.
[http://dx.doi.org/10.1039/C7RA08293B]

[31]   Wang, N.; Zhang, D.; Deng, X.; Sun, Y.; Wang, X.; Ma, P.; Song, D. A novel surface plasmon resonance biosensor based on the PDA-AgNPs-PDA-Au film sensing platform for horse IgG detection. *Spectrochim. Acta A Mol. Biomol. Spectrosc.*, **2018**, *191*, 290-295.
[http://dx.doi.org/10.1016/j.saa.2017.10.039] [PMID: 29054067]

[32]   Mota, D.R.; Lima, G.A.S.; Helene, G.B.; Pellosi, D.S. Tailoring Nanoparticle Morphology to Match Application: Growth under Low-Intensity Polychromatic Light Irradiation Governs the Morphology and Optical Properties of Silver Nanoparticles. *ACS Appl. Nano Mater.*, **2020**, *3*(5), 4893-4903.
[http://dx.doi.org/10.1021/acsanm.0c01078]

[33]   Karimzadeh, R.; Mansour, N. Thermo-optic nonlinear response of silver nanoparticle colloids under a low power laser irradiation at 532 nm. *Phys. Status Solidi, B Basic Res.*, **2010**, *247*(2), 365-370.
[http://dx.doi.org/10.1002/pssb.200945377]

[34]   Pugazhendi, S.; Kirubha, E.; Palanisamy, P.K.; Gopalakrishnan, R. Synthesis and characterization of silver nanoparticles from Alpinia calcarata by Green approach and its applications in bactericidal and nonlinear optics. *Appl. Surf. Sci.*, **2015**, *357*, 1801-1808.
[http://dx.doi.org/10.1016/j.apsusc.2015.09.237]

[35]   Zhang, R.; Liu, Y.; Kong, L.; Xu, X-Y. Regulation of optical properties for fluorescent triphenylamine-silver hybrid based on SPR effect. *Spectrochim. Acta A Mol. Biomol. Spectrosc.*, **2019**,

223: 117338.
[http://dx.doi.org/10.1016/j.saa.2019.117338] [PMID: 31306956]

[36]   Jiang, Z-J.; Liu, C-Y.; Sun, L-W. Catalytic properties of silver nanoparticles supported on silica spheres. *J. Phys. Chem. B,* **2005**, *109*(5), 1730-1735.
[http://dx.doi.org/10.1021/jp046032g] [PMID: 16851151]

[37]   Chandraker, S.K.; Lal, M.; Shukla, R. DNA-binding, antioxidant, H2O2 sensing and photocatalytic properties of biogenic silver nanoparticles using Ageratum conyzoides L. leaf extract. *RSC Advances,* **2019**, *9*(40), 23408-23417.
[http://dx.doi.org/10.1039/C9RA03590G]

[38]   Kusiak-Nejman, E.; Czyżewski, A.; Wanag, A.; Dubicki, M.; Sadłowski, M.; Wróbel, R.J.; Morawski, A.W. Photocatalytic oxidation of nitric oxide over AgNPs/TiO$_2$-loaded carbon fiber cloths. *J. Environ. Manage.,* **2020**, *262*: 110343.
[http://dx.doi.org/10.1016/j.jenvman.2020.110343] [PMID: 32250819]

[39]   Zhang, P.; Shao, C.; Zhang, Z.; Zhang, M.; Mu, J.; Guo, Z.; Liu, Y. In situ assembly of well-dispersed Ag nanoparticles (AgNPs) on electrospun carbon nanofibers (CNFs) for catalytic reduction of 4-nitrophenol. *Nanoscale,* **2011**, *3*(8), 3357-3363.
[http://dx.doi.org/10.1039/c1nr10405e] [PMID: 21761072]

[40]   Khan, A.U.; Yuan, Q.; Wei, Y.; Khan, Z.U.H.; Tahir, K.; Khan, S.U.; Ahmad, A.; Khan, S.; Nazir, S.; Khan, F.U. Ultra-efficient photocatalytic deprivation of methylene blue and biological activities of biogenic silver nanoparticles. *J. Photochem. Photobiol. B,* **2016**, *159*, 49-58.
[http://dx.doi.org/10.1016/j.jphotobiol.2016.03.017] [PMID: 27016719]

[41]   Jishma, P.; Thomas, R.; Narayanan, R.; Radhakrishnan, E.K. Exploration of photocatalytic properties of microbially designed silver nanoparticles on Victoria blue B. *Bioprocess Biosyst. Eng.,* **2016**, *39*(7), 1033-1040.
[http://dx.doi.org/10.1007/s00449-016-1581-1] [PMID: 26975321]

[42]   Elemike, E.E.; Onwudiwe, D.C.; Mkhize, Z. Eco-friendly synthesis of AgNPs using Verbascum thapsus extract and its photocatalytic activity. *Mater. Lett.,* **2016**, *185*, 452-455.
[http://dx.doi.org/10.1016/j.matlet.2016.09.026]

[43]   Khan, Z.U.H.; Khan, A.; Shah, A.; Wan, P.; Chen, Y.; Khan, G.M.; Khan, A.U.; Tahir, K.; Muhammad, N.; Khan, H.U. Enhanced photocatalytic and electrocatalytic applications of green synthesized silver nanoparticles. *J. Mol. Liq.,* **2016**, *220*, 248-257.
[http://dx.doi.org/10.1016/j.molliq.2016.04.082]

[44]   Zou, M.L.; Du, M.L.; Zhu, H.; Xu, C.S.; Fu, F.Q. Green synthesis of halloysite nanotubes supported Ag nanoparticles for photocatalytic decomposition of methylene blue. *J. Phys. D Appl. Phys.,* **2012**, *45*(32): 325302.
[http://dx.doi.org/10.1088/0022-3727/45/32/325302]

[45]   Li, J.; Li, W.; Qiang, W.; Wang, X.; Li, H.; Xu, D. A non-aggregation colorimetric assay for thrombin based on catalytic properties of silver nanoparticles. *Anal. Chim. Acta,* **2014**, *807*, 120-125.
[http://dx.doi.org/10.1016/j.aca.2013.11.011] [PMID: 24356228]

[46]   Priebe, M.; Fromm, K.M. One-pot synthesis and catalytic properties of encapsulated silver nanoparticles in silica nanocontainers. *Part. Part. Syst. Charact.,* **2014**, *31*(6), 645-651.
[http://dx.doi.org/10.1002/ppsc.201300304]

[47]   Yan, Z.; Fu, L.; Zuo, X.; Yang, H. Green assembly of stable and uniform silver nanoparticles on 2D silica nanosheets for catalytic reduction of 4-nitrophenol. *Appl. Catal. B,* **2018**, *226*, 23-30.
[http://dx.doi.org/10.1016/j.apcatb.2017.12.040]

[48]   Liu, K.; Qu, S.; Zhang, X.; Tan, F.; Wang, Z. Improved photovoltaic performance of silicon nanowire/organic hybrid solar cells by incorporating silver nanoparticles. *Nanoscale Res. Lett.,* **2013**, *8*(1), 88.
[http://dx.doi.org/10.1186/1556-276X-8-88] [PMID: 23418988]

[49]   Ramadan, R.; Silvan, M.M.; Palma, R.J.M. Hybrid porous silicon/ silver nanostructures for the development of enhanced photovoltaic devices. *J. Mater. Sci.,* **2020**, *55*(13), 5458-5470.
[http://dx.doi.org/10.1007/s10853-020-04394-z]

[50]   Joshi, D.N.; Ilairaja, P.; Sudakar, C.; Prasath, R.A. Facile one-pot synthesis of multi-shaped silver nanoparticles with tunable ultra-broadband absorption for efficient light harvesting in dye-sensitized solar cells. *Sol. Energy Mater. Sol. Cells,* **2018**, *185*, 104-110.
[http://dx.doi.org/10.1016/j.solmat.2018.05.018]

[51]   Singh, M.; Prasher, P.; Kim, J.H. Solution processed silver-nanowire/zinc oxide based transparent conductive electrode for efficient photovoltaic performance. *Nano Struc. Nano Obj.,* **2018**, *16*, 151-155.
[http://dx.doi.org/10.1016/j.nanoso.2018.05.009]

[52]   Liu, X-H.; Hou, L-X.; Wang, J-F.; Liu, B.; Yu, Z-S.; Ma, L-Q.; Yang, S-P.; Fu, G-S. FU, G-S. Plasmonic-enhanced polymer solar cells with high efficiency by addition of silver nanoparticles of different sizes in different layers. *Sol. Energy,* **2014**, *110*, 627-635.
[http://dx.doi.org/10.1016/j.solener.2014.06.019]

[53]   Joshi, D.N.; Mandal, S.; Kothandraman, R.; Prasath, R.A. Efficient light harvesting in dye sensitized solar cells using broadband surface plasmon resonance of silver nanoparticles with varied shapes and sizes. *Mater. Lett.,* **2017**, *193*, 288-291.
[http://dx.doi.org/10.1016/j.matlet.2017.02.008]

[54]   Ghadari, R.; Sabri, A.; Saei, P-S.; Kong, F-T.; Marques, H.M. Phthalocyanine-silver nanoparticle structures for plasmon-enhanced dye-sensitized solar cells. *Sol. Energy,* **2020**, *198*, 283-294.
[http://dx.doi.org/10.1016/j.solener.2020.01.053]

[55]   Seoudi, R.; Althagafi, H.A. Dependence of Copper Phthalocyanine Photovoltaic Thin Film on the Sizes of Silver Nanoparticles. *Silicon,* **2018**, *10*(5), 2165-2171.
[http://dx.doi.org/10.1007/s12633-017-9748-1]

[56]   Zhang, Q.; Qin, W-J.; Cao, H-Q.; Yang, L-Y.; Zhang, F-L.; Yin, S-G. Effects of the position of silver nanoprisms on the performance of organic solar cells. *Optoelectron. Lett.,* **2014**, *10*(4), 253-257.
[http://dx.doi.org/10.1007/s11801-014-4041-7]

[57]   Su, Y-A.; Lin, W-C.; Wang, H-J.; Lee, W-H.; Lee, R-H.; Dai, S.A.; Hsieh, C-F.; Jeng, R-J. Enhanced photovoltaic performance of inverted polymer solar cells by incorporating graphene nanosheet/AgNPs nanohybrids. *RSC Advances,* **2015**, *5*(32), 25192-25203.
[http://dx.doi.org/10.1039/C4RA16855K]

[58]   Metzman, J.S.; Khan, A.U.; Magill, B.A.; Khodaparast, G.A.; Heflin, J.R.; Liu, G. Critical Role of Polystyrene Layer on Plasmonic Silver Nanoplates in Organic Photovoltaics. *ACS Appl. Energy Mater.,* **2019**, *2*(4), 2475-2485.
[http://dx.doi.org/10.1021/acsaem.8b01860]

[59]   Higgins, M.; Ely, F.; Nome, R.C.; Nome, R.A.; Santos, D.P.; Choi, H.; Nam, S.; Lopez, M.Q. Enhanced reproducibility of planar perovskite solar cells by fullerene doping with silver nanoparticles. *J. Appl. Phys.,* **2018**, *124*(6): 065306.
[http://dx.doi.org/10.1063/1.5036643]

[60]   Kazmi, S.A.; Hameed, S.; Azam, A. Efficiency enhancement in dye-sensitized solar cells using silver nanoparticles and TiCl$_4$. *Energy Sources A Recovery Util. Environ. Effects,* **2017**, *39*(1), 67-74.
[http://dx.doi.org/10.1080/15567036.2016.1205682]

[61]   Sreeja, S.; Pesala, B. Efficiency Enhancement of Betanin–Chlorophyll Cosensitized Natural Pigment Solar Cells Using Plasmonic Effect of Silver Nanoparticles. *IEEE Int. J. Photovolt.,* **2020**, *10*(1), 124-134.
[http://dx.doi.org/10.1109/JPHOTOV.2019.2953399]

[62]   Yu, S-J.; Yin, Y-G.; Liu, J-F. Silver nanoparticles in the environment. *Environ. Sci. Process. Impacts,*

2013, *15*(1), 78-92.
[http://dx.doi.org/10.1039/C2EM30595J] [PMID: 24592429]

[63]   Ferdous, Z.; Nemmar, A. Health Impact of Silver Nanoparticles: A Review of the Biodistribution and Toxicity Following Various Routes of Exposure. *Int. J. Mol. Sci.,* **2020**, *21*(7), 2375.
[http://dx.doi.org/10.3390/ijms21072375] [PMID: 32235542]

[64]   Levard, C.; Hotze, E.M.; Lowry, G.V.; Brown, G.E., Jr Environmental transformations of silver nanoparticles: impact on stability and toxicity. *Environ. Sci. Technol.,* **2012**, *46*(13), 6900-6914.
[http://dx.doi.org/10.1021/es2037405] [PMID: 22339502]

[65]   Sekeryan, S.T.; Hicks, A.L. Understanding the potential environmental benefits of nanosilver enabled consumer products. *NanoImpact.,* **2020**, *16*, Article 100183.

[66]   Mao, B-H.; Chen, Z-Y.; Wang, Y-J.; Yan, S-J. Silver nanoparticles have lethal and sublethal adverse effects on development and longevity by inducing ROS-mediated stress responses. *Sci. Rep.,* **2018**, *8*(1), 2445.
[http://dx.doi.org/10.1038/s41598-018-20728-z] [PMID: 29402973]

CHAPTER 2

Synthesis of Silver Nanoparticles

Abstract: The synthesis of AgNPs occurs by traditional chemical synthesis, which requires reducing agents to convert the ionic silver to metallic form and a capping agent that stabilizes the colloidal suspension of AgNPs. While the green synthesis inculcates the use of natural agents and their extract or dry biomass that play a dual role of reducing and capping agents. Mainly, the alkaloids, natural products, terpenes, polysaccharides, and coumarins present in the biomass or their extract serve as the reducing/ capping agent for the synthesis of AgNPs *via* the green route. The present chapter discusses the modes of synthesis of AgNPs by the green routes.

Keywords: AgNPs morphology, Capping agent, Chemical synthesis, Green synthesis, Microbial synthesis.

1. CHEMICAL SYNTHESIS

The chemical synthesis of AgNPs starts with the reduction of Ag(I) solution in the presence of a suitable reducing agent to obtain Ag(0) nuclei aggregates in the colloidal solution. The metallic silver nuclei in the presence of a suitable stabilizing agent coalease to form clusters in the colloidal solution that generate colloidal stabilized AgNPs. Further modifications on AgNPs for fine-tuning of their surface and morphology occur in the presence of chemical and physical modifications in the colloidal state. Fig. (**1**) illustrates the formation of surface functionalized AgNPs. Mainly the chemical reducing agents such as $NaBH_4$ and $LiAlH_4$ represent the most commonly used reagents for the conversion of silver ions in solution phase to metallic silver. The stabilizing agents used in chemical synthesis mainly include a synthetic polymer that prevents the coalease of AgNPs in colloidal phase and keeps them dispersed in the solution itself. However, in the present era, more emphasis occurs on the green methods of AgNPs synthesis owing to environmental sustainability and low harm to the ecosystem.

1.1. Green Synthesis

Fig. (**2**) highlights the various sources for the green synthesis of AgNPs. Mainly, the various phytochemicals, natural products, macromolecules, tannins, lignins, coumarins, terpenes, and heterocyclic or spirocyclic components present in the sources mentioned in Fig. (**2**) assist in the AgNPs synthesis. These agents serve the dual role of being a reducing and coupling agent and provide a high yield of AgNPs. Tables **1-3** depict the green synthesis of AgNPs by natural sources in the form of microbes and agrowaste.

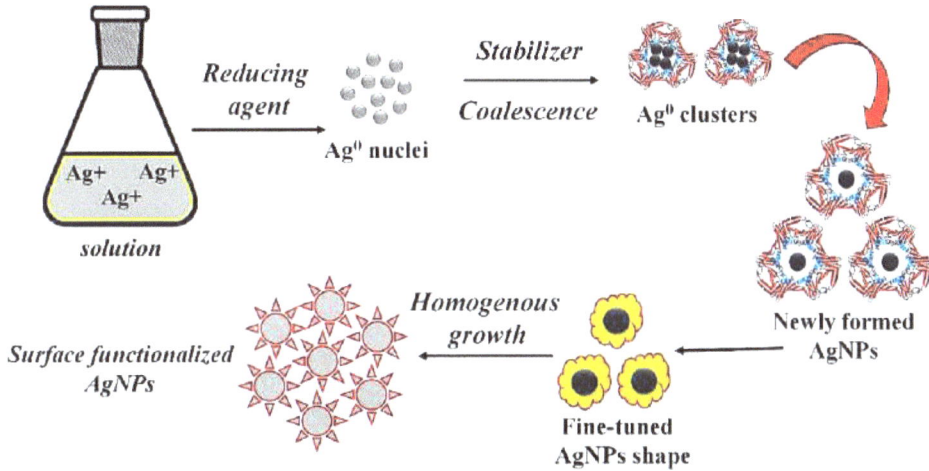

Fig. (1). Sequence of steps for the synthesis of surface functionalized AgNPs.

Fig. (2). Sources for the green synthesis of AgNPs.

Table 1. Synthesis by bacteria.

Bacteria	AgNPs Characteristics	Reducing Agent	Refs.
Extracellular synthesis by *Pseudomonas antarctica, Pseudomonas proteolytica, Pseudomonas meridiana, Arthrobacter kerguelensis and Arthrobacter gangotriensis and two mesophilic bacteria Bacillus indicus and Bacillus cecembensis*	Spherical shape, average diameter in the range 6-13 nm	Cell free culture supernatant	[1]
Endophytic bacterium was isolated from the plant red fountain grass (*Pennisetum setaceum*)	Monodispersed spherical shape, average diameter in the range 83-176 nm	Bacterial wet biomass	[2]
Acidophilic actinobacterial SH11 strain isolated from pine forest soil	Polydispersed spherical shape, mean diameter 13.2 nm	Cell free culture supernatant	[3]
Endophytic Bacterium *Pantoea ananatis*	Spherical shape, average diameter in the range 8.06-91.32 nm	Cell free culture supernatant	[4]
Bacillus brevis (NCIM 2533)	Spherical shape, average diameter in the range 41-68 nm	Bacterial cell filtrate	[5]
Extracellular biosynthesis with *Bacillus methylotrophicus* DC3 strain	Spherical shape, average diameter in the range 10-30 nm	Bacterial dry biomass	[6]
Extracellular synthesis by Soil bacteria *Cupriavidus sp*	Spherical, crystalline shape, average diameter in the range 10-50 nm	Cell free supernatant	[7]
Bacterial exopolysaccharide of *Leuconostoc lactis*	Spherical shape, average diameter 35 nm	Partially purified exopolysaccharide	[8]
Extracellular synthesis by *Pseudomonas* sp. THG-LS1.4	Irregular shape, average diameter in the range 10-40 nm	Cell free supernatant	[9]
Streptomyces xinghaiensis OF1 strain	Spherical, polydispersed nanoparticles, average diameter in the range 5-20 nm	Cell free supernatant	[10]
Sphingobium sp MAH11	Spherical nanoparticles, average diameter in the range 7-22 nm	Cell free supernatant	[11]
Lactobacillus sp. LCM5	Spherical nanoparticles, average diameter in the range 3-35 nm	Culture filtrate of bacteria	[12]
Shewanella sp. ARY1	Spherical nanoparticles, mean diameter 38 nm	Culture supernatant of bacteria	[13]
Bacillus cereus	Spherical nanoparticles, average diameter in the range 10-30 nm	Cell free supernatant	[14]

Bacteria	AgNPs Characteristics	Reducing Agent	Refs.
Extracellular synthesis by *Thermophilic Bacillus Sp.* AZ1	Spherical nanoparticles, average diameter in the range 7-31 nm	Cell free supernatant	[15]
Bacillus cereus SZT1	Spherical nanoparticles, average diameter in the range 18-39 nm	Culture supernatant of bacteria	[16]
Pseudoduganella eburnea MAHUQ-39	Spherical nanoparticles, average diameter in the range 8-24 nm	Culture supernatant of bacteria	[17]
Endophytic bacteria *Bacillus siamensis* strain C1 isolated from *Coriandrum sativum*	Spherical nanoparticles, average diameter in the range 25-50 nm	Cell free supernatant	[18]
Extracellular synthesis by *Exiguobacterium* sp. KNU1	Spherical nanoparticles, average diameter in the range 5-50 nm	Extracellular bacterial enzymes	[19]
Bacillus safensis LAU 13	Spherical nanoparticles, average diameter in the range 5-30 nm	Bacterial keratinase	[20]
Citrobacter spp. MS5	Spherical nanoparticles, average diameter in the range 5-15 nm	Bacterial culture supernatant	[21]
Bacillus amyloliquefaciens	Circular and triangular crystalline AgNPs, mean diameter 14.6 nm	Cell free bacterial extract	[22]
Pseudomonas aeruginosa JQ989348	Anisotropic crystalline structure, two types of nanoparticles with size 13 nm and 76 nm	Cell free extract	[23]
Extracellular synthesis from *Phenerochaete chrysosporium* (MTCC-787)	Spherical and oval shaped nanoparticles, average diameter 34 nm and 90 nm	Cell free extract	[24]
Pseudomonas rhodesiae	Crystalline, 20-100 nm	Culture supernatant	[25]
Mixed bacterial culture	Spherical, crystalline nanoparticles, mean diameter 7±3 nm	Polyhydroxyalkanoate	[26]
Acinetobacter sp. GWRFHc45	Monodispersed nanoparticles, average diameter 20 nm	Cell free supernatant	[27]
Acinetobacter sp.	Spherical polydispersed nanoparticles, average diameter 50 nm	Lignin peroxidase	[28]
Psudomonas aeruginosa	Spherical nanoparticles, average diameter 30-70 nm	Cell free supernatant	[29]
Acidophilic strain of *Actinobacteria* isolated from the of Picea sitchensis forest soil	Crystalline nanoparticles, average diameter in the range 4-45 nm	Cell free supernatant	[30]
Acinetobacter calcoaceticus	Polydispersed nanoparticles, average diameter in the range 10-60 nm	Cell free supernatant	[31]

(Table 1) cont.....

Bacteria	AgNPs Characteristics	Reducing Agent	Refs.
Lactobacillus brevis NM101-1	Spherical shaped nanoparticles, average diameter in the range 11-25 nm	Bacterial exopolyaccharides	[32]
Mesoflavibacter zeaxanthinifaciens	Spherical shaped nanoparticles, average diameter in the range 35 nm	Bacterial exopolyaccharides	[33]
Thermophilic strain (Ts-1) of *Bacillus amyloliquefaciens*	Spherical shaped nanoparticles, average diameter in the range 10-50 nm	Bacterial exopolyaccharides	[34]
Lactobacillus sp.	Spherical shaped nanoparticles, average diameter in the range 5-60 nm	Bacterial exopolyaccharides	[35]
Bacillus amyloliquefaciens	Spherical shaped nanoparticles, average diameter in the range 4-20 nm	Bacterial exopolyaccharides	[36]

Table 2. Synthesis by fungi.

Fungi	AgNPs Characteristics	Reducing Agent	Refs.
Aspergillus niger L3 (NEA) and *Trichoderma longibrachiatum* L2 (TEA)	Spherical shaped nanoparticles, average diameter in the range 15.21-77.49 nm	Fungal xylanases	[37]
Endophyticus fungus *Botryosphaeria rhodina*	Spherical shaped nanoparticles, average diameter in the range 2-50 nm	Secondary metabolites	[38]
Aspergillus terreus	Polydispersed nanoparticles, average diameter in the range 1-20 nm	NADPH-dependent reductase	[39]
Aspergillus clavatus	Polydispersed spherical and hexagonal nanoparticles, average diameter in the range 10-25 nm	Fungal conidiophore	[40]
Marine endophytic fungus *Cladosporium cladosporioides*	Spherical shape, average diameter 30-60 nm	NADPH-dependent reductase	[41]
Medicinal fungus Cs-HK1	Spherical shape, mean diameter 50, narrow size distribution	Water soluble pxopolysaccharide EPSI	[42]
White rot fungi	Spherical shape, average diameter 15-25 nm	Culture filtrate extracts	[43]
Aspergillus oryzae (MTCC No. 1846)	Spherical shape with variable sizes	Fungal cell filtrate	[44]
Endophytic fungi *Raphanus sativus*	Spherical shape, average diameter 4-30 nm	Supernatant	[45]
Endophytic fungi *Aspergillus versicolor* ENT7	Spherical shape, average diameter 3-40 nm	Fungal isolate	[46]

(Table 2) cont.....

Fungi	AgNPs Characteristics	Reducing Agent	Refs.
Ligninolytic fungi *Trametes trogii*	Spherical, core shell, and ellipsoid nanoparticles	Fungal extract	[47]
Cladosporium halotolerans	Variable shapes and sizes	Fungal extract	[48]
Tinospora cordifolia	Spherical shape, average diameter 25-35 nm	Endophyte fungus isolate	[49]
White rot fungi *Pycnoporus sanguineus*	Spherical shape, average diameter 52.8-103.3 nm	Proteins from mycelial surface	[50]
Aspergillus fumigatus MA	Spherical shape, crystalline nanoparticles, average diameter 3-20 nm	Fungal isolate	[51]
Trichoderma reesei	Spherical shape, crystalline nanoparticles, average diameter 5-50 nm	Fungal mycelium	[52]
Entomopathogenic fungus *(Beauveria bassiana)*	Triangular, circular and hexagon shaped nanoparticles, average diameter in the range 10-50 nm	Fungal supernatant	[53]
Penicillium oxalicum	Spherical shape, average diameter 4 nm	Fungal cell filtrate	[54]
Trichoderma viride	Spherical shape, average diameter 4-16 nm	Fungal cell filtrate	[55]
Cladosporium sp.	Spherical shape, mean diameter 24 nm	NADPH-dependent reductase	[56]
Cochliobolus lunatus	Spherical shape, average diameter 3-21 nm	Fungal cell filtrate	[57]
Amylomyces rouxii strain KSU-09	Spherical, monodispersed nanoparticles, average diameter 5-27 nm	Mycelia free fungal extract	[58]
Endophytic fungi *Cryptosporiopsis ericae* PS4	Spherical shape, average diameter 5.5 nm	Dried fungal biomass	[59]
Endophytic fungi *Colletotrichum incarnatum* DM16.3	Highly aggregated, spherical, crystalline nanoparticles, average diameter 5-25 nm	Fungal cell filtrate	[60]
Endophytic fungus *Penicillium polonicum*	Spherical shape, average diameter 10-15 nm	Fungal biomass	[61]
Aspergillus tamarii	Spherical shape, mean diameter 40 nm	Fungal biomass	[62]
Endophytic fungus *Talaromyces purpureogenus*	Triangle-shaped nanoparticles, mean diameter 25 nm	Wet fungal biomass	[63]
Sclerotinia sclerotiorum MTCC 8785	Spherical shape, mean diameter 10 nm	Wet fungal biomass	[64]

(Table 2) cont.....

Fungi	AgNPs Characteristics	Reducing Agent	Refs.
Aspergillus terreus HA1N and *Penicillium expansum* HA2N	Spherical shape, mean diameter 10-25 nm	Fungal biomass	[65]
Oropharyngeal *Candida glabrata* Isolates	Well-dispersed nanoparticles, average diameter in the range 2-15 nm	Supernatant	[66]
Fusarium solani	Spherical shape, mean diameter 10 nm	Supernatant	[67]
Filamentous fungi *Penciillium Citreonigum Dierck* and *Scopulaniopsos brumptii Salvanet-Duval*	Monodispersed nanoparticles, average diameter in the range 3-23.2 nm	Fungal biomass	[68]
Cryphonectria sp.	Monodispersed nanoparticles, average diameter in the range 30-70 nm	Fungal cell filtrate	[69]
Endophytic Fungi *Penicillium sp.*	Well-dispersed, spherical nanoparticles, average diameter 25-30 nm	Wet fungal biomass	[70]
Neurospora intermedia	Spherical nanoparticles, mean diameter 24 nm	Wet fungal biomass	[71]
Fusarium oxysporum	Spherical nanoparticles, mean diameter 20.7 nm	Enzymatic generation of extracellular superoxide	[72]
Duddingtonia flagrans	Spherical nanoparticles, average diameter in the range 1.18-14.51 nm	Fungal cell isolate	[73]

Table 3. Synthesis by algae.

Algae	AgNPs Characteristics	Reducing Agent	Refs.
Brown marine algae *Cystophora moniliformis*	Spherical nanoparticles, mean diameter 75 nm	Algae extract	[74]
Green algae *Caulerpa serrulata*	Spherical nanoparticles, mean diameter 10 nm	Algae extract	[75]
Marine algae *Caulerpa racemos*	Crystalline, face-centered cubic arrangement of nanoparticles, average diameter in the range 5-25 nm	Dry algae biomass	[76]
Seaweeds *Ulva fasciata, Corallina elongate, Gelidium crinale, Laurencia obtusa, Cystoseira myrica* and *Turbinaria turbinata*	Spherical nanoparticles, average diameter in the range 8-16 nm	Algae extract	[77]
Marine Algae *Ecklonia cava*	Spherical nanoparticles, mean diameter 43 nm	Phenolic compounds in the aqueous extract of algae	[78]
Marine red alga *Laurencia catarinensis*	Spherical nanoparticles, average diameter in the range 39.41-77.71 nm	Powdered algae extract	[79]

(Table 3) cont.....

Algae	AgNPs Characteristics	Reducing Agent	Refs.
Polysiphonia algae	Spherical nanoparticles, average diameter in the range 5-25 nm	Cell-free extract	[80]
Fresh water green alga *Pithophora oedogonia*	Spherical nanoparticles, mean diameter 34.03 nm	Algae extract	[81]
Green algae *Ulva armoricana* sp.	Spherical, crystalline nanoparticles, mean diameter 33 nm	Ulvan obtained from algae biomass	[82]
Marine macro-algae *Pterocladia capillacae, Jania rubins, Ulva faciata, and Colpmenia sinusa*	Spherical, polydispersed nanoparticles, average diameter in the range 7-20 nm	Polysaccharides present in the algae extract	[83]
Marine green algae *Caulerpa racemosa*	Distorted spherical nanoparticles, mean diameter 25 nm	Phytoconstituents in algae biomass	[84]
Marine red algae *Gelidium corneum*	Crystalline nanoparticles with centric cubic geometry, average diameter in the range 20-50 nm	Algae extract	[85]
Red algae *Amphiroa rigida*	Face centered cubic crystalline nanoparticles, mean diameter 25 nm	Functional groups on the polysaccharides present in algae biomass	[86]
Green alga *Chlorella ellipsoidea*	Spherical nanoparticles, mean diameter 220 nm	Dried algae biomass	[87]
Macro-Algae *Gracilaria edulis*	Spherical nanoparticles, average diameter in the range 55-99 nm	Aqueous extracts of algae	[88]
Marine brown alga *Padina pavonia*	Variable shapes, including spherical, rectangle, triangular, hexagonal, and polyhedral shapes, average diameter in the range 49.58-86.37 nm	Powdered algae biomass	[89]
Marine Red Algae *Acanthophora specifera*	Cubic nanoparticles, average diameter in the range 33-81 nm	Powdered algae biomass	[90]
Green algae *Noctiluca scintillans*	Spherical nanoparticles, mean diameter 4.5 nm	Algae extract	[91]
Marine algae *Porphyra yezoensis*	Spherical, well-dispersed nanoparticles, mean diameter 7.2 nm	Green protein R-phycoerythrin	[92]
Red algae *Laurencia aldingensis* and *Laurenciella* sp.	Spherical nanoparticles, mean diameter 15.4 nm	Aqueous extract of algae	[93]
Spirulina microalgae	Spherical, well-dispersed nanoparticles, average diameter 5-50 nm	Algae extract	[94]
Green algae *P. kessleri*	Spherical nanoparticles, average diameter 15-35 nm	Algae extract	[95]
Green alga *Neochloris oleoabundans*	Quasi-spherical nanoparticles, mean diameter 16.63 nm	Algae cell extract	[96]

(Table 3) cont.....

Algae	AgNPs Characteristics	Reducing Agent	Refs.
Chlamydomonas reinhardtii	Variable sizes	Algae cell extract	[97]
Red algae *Gracilaria firma*	Variable sizes	Algae cell extract	[98]
Gracilaria birdiae	Spherical nanoparticles, average diameter in the range 20.2-94.9 nm	Polysaccharides present in algae cell extract	[99]
Chlorella vulgaris	Spherical shaped, crystalline nanoparticles, average diameter in the range 55.06-61.89 nm	Algae extract	[100]
Marine algae *Sargassum wightii* and *Valonopsis pachynema*	Spherical nanoparticles, average diameter in the range 30-70 nm	Methanolic extract of algae	[101]
Marine macroalgae *Padina* sp.	Spherical nanoparticles, average diameter in the range 30-70 nm	Algae extract	[102]
Cladophora glomerata	Spherical nanoparticles, average diameter in the range 8-11 nm	Algae extract	[103]
Marine microalgae *Trichodesmium erythraeum*	Spherical nanoparticles, cubic shaped, mean diameter in the range 26.5 nm	Algae extract	[104]
Desmodesmus sp.	Spherical nanoparticles, average diameter in the range 15-30 nm	Chlorophyll content in algae	[105]
Gracilaria corticata, G. edulis, Hypnea musciformis and Spyridia hypnoides	Spherical nanoparticles, average diameter in the range 37-54 nm	Aqueous extract of algae	[106]
Codium capitatum P.C. Silva	Spherical nanoparticles, average diameter in the range 3-44 nm	Dried extract of algae	[107]
Chlamydomonas reinhardtii	Spherical ultrafine nanoparticles, average diameter <10 nm	Extracellular polymeric substances	[108]
Green Alga *Botryococcus braunii*	Spherical nanoparticles, average diameter in the range 58-89 nm	Algae extract	[109]
Laminaria japonica	Spherical nanoparticles, mean diameter 20 nm	Algae extract	[110]

Table 4. Synthesis with agro-waste.

Agro-waste	AgNPs Characteristics	Reducing Agent/ Waste Material	Refs.
Annona squamosa	Irregular spherical shaped nanoparticles, mean diameter 35 nm	Peel extract	[111]
Cocos nucifera	Spherical shaped nanoparticles, average diameter in the range14.2-22.96 nm	Shell	[112]
Poa annua	Spherical shaped nanoparticles, mean diameter 36.66 nm	Extract	[113]

(Table 4) cont.....

Agro-waste	AgNPs Characteristics	Reducing Agent/ Waste Material	Refs.
Citrus sinensis	Spherical shaped nanoparticles, mean diameter 23.81 nm	Peels	[114]
Cocos nucifera	Spherical shaped nanoparticles	Outer shell fibre	[115]
Tectona grandis Linn.	Spherical shaped nanoparticles, mean diameter 28 nm	Leaf extract	[116]
P. Americana, Beta vulgaris, Arachis hypogaea	Spherical shaped nanoparticles	Peel and shell	[117]
Cola nitida	Spherical shaped nanoparticles, average diameter in the range12-80 nm	Pod extract	[118]
Vitis vinifera	Spherical shaped nanoparticles, mean diameter 33 nm	Pomace	[119]
Ananas comosus	Spherical shaped nanoparticles	Peel extract	[120]
Garcinia Mangostana and *Nephelium lappaceum*	Spherical shaped nanoparticles, average diameter in the range 12.1-31.3 nm	Water extract	[121]
Citrullus lanatus	Spherical shaped nanoparticles, mean diameter 17.96 nm	Fruit rind extract	[122]
Plukenetia volubilis	Spherical shaped nanoparticles, mean diameter 7.2 nm	Shell biomass	[123]
Carica Papaya	Spherical shaped nanoparticles, mean diameter 28 nm	Peel extract	[124]
Citrus reticulum	Spherical/ oval shaped nanoparticles, average diameter in the range12-80 nm	Waste peels	[125]
Camellia sinensis	Ultrafine spherical nanoparticles, average diameter <20 nm	Stem waste	[126]
Physalis peruviana L.	Spherical shaped nanoparticles, average diameter in the range 25-55 nm	Non-edible Accrescent Fruiting Calyx	[127]
Oryza sativa	Variable shapes and sizes	Husk metabolites including alcohols, sugars and sugar derivatives, amino acids, organic acids, fatty acids, and medicinal compounds, including xylitol, sinapyl alcohol, levoglucosan, glucose, carbamoyl-aspartic acid, citrulline, maleimide, phytol, and acetanilide	[128]
Saccharum officinarum	Spherical shaped nanoparticles, mean diameter 16.9 nm	Leaves extract	[129]

(Table 4) cont.....

Agro-waste	AgNPs Characteristics	Reducing Agent/ Waste Material	Refs.
Vitis vinifera	Spherical shaped nanoparticles, mean diameter 54.8 nm	Seed extract	[130]
Punica granatum	Spherical shaped nanoparticles, average diameter in the range 20-50 nm	Peel extract	[131]
Triticum aestivum	Polydispersed spherical nanoparticles, average diameter in the range 20-45 nm	Xylans in bran	[132]
Theobroma cacao	Spherical nanoparticles, average diameter in the range 4-32 nm	Pod husk	[133]
Punica granatum	Spherical nanoparticles, average diameter in the range 20-40 nm	Peel extract	[134]
Prunus dulcis	Spherical nanoparticles, average diameter in the range 5-15 nm	Waste shell	[135]
Melia azedarach	Spherical shaped nanoparticles, mean diameter 20 nm	Bark extract	[136]
Durio zibethinus	Spherical shaped nanoparticles, mean diameter 11.4 nm	Mesocarp and endocarp	[137]
Lantana camara	Spherical shaped nanoparticles, mean diameter 75.2 nm	Berries	[138]
Crocus sativus L.	Spherical shaped nanoparticles, average diameter in the range 12-20 nm	Aqueous extract	[139]
Musa paradisiaca	Spherical shaped crystalline nanoparticles, mean diameter 34 nm	Peel extract	[140]
Citrus X sinensis	Spherical shaped nanoparticles, average diameter in the range 47-53 nm	Peel Extract	[141]
Saccharum officinarum	Spherical shaped crystalline nanoparticles, mean diameter 28.2 nm	Waste leaves	[142]
Cocos nucifera	Spherical shaped nanoparticles, average diameter in the range 10-70 nm	Oil cake extract	[143]
Zea mays	Various shapes and sizes	Stalk	[144]
Oryza sativa	Spherical shaped ultrafine nanoparticles, average diameter <15 nm	Husk	[145]
Oryza sativa	Spherical shaped nanoparticles, mean diameter 20 nm	Husk, husk ash	[146]
Oryza sativa	Spherical shaped nanoparticles, mean diameter 19 nm	Phenolic groups of husk	[147]
Oryza sativa	Spherical shaped nanoparticles, average diameter in the range 1-100 nm	Husk	[148]

CONCLUSION

The green synthesis of AgNPs *via* microbes and agro-biomass serves as a sustainable approach for the effective utilization of the latter for the generation of value added, functional AgNPs. The green synthesized AgNPs discourage the use of chemical reagents or hazardous solvents for the synthesis that prove detrimental to environmental health. However, the ultimate fate of green synthesized AgNPs could serve a direct exposure to the ecosystem where the AgNPs on losing their surface capping freely interact with the various realms of the ecosystem. Such AgNPs eventually become a part of various trophic levels and closely integrate with the food chains and food webs. Therefore, the utilization of AgNPs *via* the green route still necessitates the consideration of their ultimate fate in the environment for mitigating the lost lasting effects.

REFERENCES

[1]   Shivaji, S.; Madhu, S.; Singh, S. Extracellular synthesis of antibacterial silver nanoparticles using psychrophilic bacteria. *Process Biochem.,* **2011**, *46*(9), 1800-1807.
[http://dx.doi.org/10.1016/j.procbio.2011.06.008]

[2]   Shariq Ahmed, M.; Soundhararajan, R.; Akther, T.; Kashif, M.; Khan, J.; Waseem, M.; Srinivasan, H. Biogenic AgNPs synthesized *via* endophytic bacteria and its biological applications. *Environ. Sci. Pollut. Res. Int.,* **2019**, *26*(26), 26939-26946.
[http://dx.doi.org/10.1007/s11356-019-05869-6] [PMID: 31309423]

[3]   Wypij, M.; Golinska, P.; Dahm, H.; Rai, M. Actinobacterial-mediated synthesis of silver nanoparticles and their activity against pathogenic bacteria. *IET Nanobiotechnol.,* **2017**, *11*(3), 336-342.
[http://dx.doi.org/10.1049/iet-nbt.2016.0112] [PMID: 28476992]

[4]   Monowar, T.; Rahman, M.S.; Bhore, S.J.; Raju, G.; Sathasivam, K.V. Silver Nanoparticles Synthesized by Using the Endophytic Bacterium *Pantoea ananatis* are Promising Antimicrobial Agents against Multidrug Resistant Bacteria. *Molecules,* **2018**, *23*(12), 3220.
[http://dx.doi.org/10.3390/molecules23123220] [PMID: 30563220]

[5]   Saravanan, M.; Arokiyaraj, S.; Lakshmi, T.; Pugazhendhi, A. Synthesis of silver nanoparticles from Phenerochaete chrysosporium (MTCC-787) and their antibacterial activity against human pathogenic bacteria. *Microb. Pathog.,* **2018**, *117*, 68-72.
[http://dx.doi.org/10.1016/j.micpath.2018.02.008] [PMID: 29427709]

[6]   Wang, C.; Kim, Y.J.; Singh, P.; Mathiyalagan, R.; Jin, Y.; Yang, D.C. Green synthesis of silver nanoparticles by *Bacillus methylotrophicus*, and their antimicrobial activity. *Artif. Cells Nanomed. Biotechnol.,* **2016**, *44*(4), 1127-1132.
[PMID: 25749281]

[7]   Ameen, F.; AlYahya, S.; Govarthanan, S.; Aljahdali, N.; Al-Enazi, N.; Alsamhary, K.; Alshehri, W.A.; Alwakeel, S.S.; Alharbi, S.S. Soil bacteria *Cupriavidus sp.* mediates the extracellular synthesis of antibacterial silver nanoparticles. *J. Mol. Struct.,* **2020**, *1202*: 127233.
[http://dx.doi.org/10.1016/j.molstruc.2019.127233]

[8]   Saravanan, C.; Rajesh, R.; Kaviarasan, T.; Muthukumar, K.; Kavitake, D.; Shetty, P.H. Synthesis of silver nanoparticles using bacterial exopolysaccharide and its application for degradation of azo-dyes. *Biotechnol. Rep. (Amst.),* **2017**, *15*, 33-40.
[http://dx.doi.org/10.1016/j.btre.2017.02.006] [PMID: 28664148]

[9]   Singh, H.; Du, J.; Singh, P.; Yi, T.H. Extracellular synthesis of silver nanoparticles by *Pseudomonas* sp. THG-LS1.4 and their antimicrobial application. *J. Pharm. Anal.,* **2018**, *8*(4), 258-264.

[http://dx.doi.org/10.1016/j.jpha.2018.04.004] [PMID: 30140490]

[10]   Wypij, M.; Czarnecka, J.; Świecimska, M.; Dahm, H.; Rai, M.; Golinska, P. Synthesis, characterization and evaluation of antimicrobial and cytotoxic activities of biogenic silver nanoparticles synthesized from *Streptomyces xinghaiensis* OF1 strain. *World J. Microbiol. Biotechnol.,* **2018**, *34*(2), 23.
[http://dx.doi.org/10.1007/s11274-017-2406-3] [PMID: 29305718]

[11]   Akter, S.; Huq, M.A. Biologically rapid synthesis of silver nanoparticles by *Sphingobium* sp. MAH-11^T and their antibacterial activity and mechanisms investigation against drug-resistant pathogenic microbes. *Artif. Cells Nanomed. Biotechnol.,* **2020**, *48*(1), 672-682.
[http://dx.doi.org/10.1080/21691401.2020.1730390] [PMID: 32075448]

[12]   Matei, A.; Matei, S.; Matei, G-M.; Cogalniceanu, G.; Cornea, C.P. Biosynthesis of silver nanoparticles mediated by culture filtrate of lactic acid bacteria, characterization and antifungal activity. *EuroBiotech J.,* **2020**, *4*(2), 97-103.
[http://dx.doi.org/10.2478/ebtj-2020-0011]

[13]   Mondal, A.H.; Yadav, D.; Ali, A.; Khan, N.; Jin, J.O.; Haq, Q.M.R. Anti-Bacterial and Anti-Candidal Activity of Silver Nanoparticles Biosynthesized Using *Citrobacter* spp. MS5 Culture Supernatant. *Biomolecules,* **2020**, *10*(6), 944.
[http://dx.doi.org/10.3390/biom10060944] [PMID: 32580522]

[14]   Prakash, A.; Sharma, S.; Ahmad, N.; Sinha, P. Synthesis of AgNPs by Bacillus cereus and their antimicrobial potential. *J. Biomater. Nanobiotechnol.,* **2011**, *2*(2), 155-161.
[http://dx.doi.org/10.4236/jbnb.2011.22020]

[15]   Deljou, A.; Goudarzi, S. Green Extracellular Synthesis of the Silver Nanoparticles Using Thermophilic *Bacillus Sp.* AZ1 and its Antimicrobial Activity Against Several Human Pathogenetic Bacteria. *Iran. J. Biotechnol.,* **2016**, *14*(2), 25-32.
[http://dx.doi.org/10.15171/ijb.1259] [PMID: 28959323]

[16]   Ahmed, T.; Shahid, M.; Noman, M.; Niazi, M.B.K.; Mahmood, F.; Manzoor, I.; Zhang, Y.; Li, B.; Yang, Y.; Yan, C.; Chen, J. Silver Nanoparticles Synthesized by Using *Bacillus cereus* SZT1 Ameliorated the Damage of Bacterial Leaf Blight Pathogen in Rice. *Pathogens,* **2020**, *9*(3), 160.
[http://dx.doi.org/10.3390/pathogens9030160] [PMID: 32110981]

[17]   Huq, M.A. Green Synthesis of Silver Nanoparticles Using *Pseudoduganella eburnea* MAHUQ-39 and Their Antimicrobial Mechanisms Investigation against Drug Resistant Human Pathogens. *Int. J. Mol. Sci.,* **2020**, *21*(4), 1510.
[http://dx.doi.org/10.3390/ijms21041510] [PMID: 32098417]

[18]   Ibrahim, E.; Fouad, H.; Zhang, M.; Qiu, W.; Yan, C.; Li, B.; Mo, J.; Chen, J. Biosynthesis of silver nanoparticles using endophytic bacteria and their role in inhibition of rice pathogenic bacteria and plant growth promotion. *RSC Advances,* **2019**, *9*(50), 29293-29299.
[http://dx.doi.org/10.1039/C9RA04246F]

[19]   Tamboli, D.P.; Lee, D.S. Mechanistic antimicrobial approach of extracellularly synthesized silver nanoparticles against gram positive and gram negative bacteria. *J. Hazard. Mater.,* **2013**, *260*, 878-884.
[http://dx.doi.org/10.1016/j.jhazmat.2013.06.003] [PMID: 23867968]

[20]   Lateef, A.; Adelere, I.A.; Kana, E.B.G.; Asafa, T.B.; Beukes, L.S. Green synthesis of silver nanoparticles using keratinase obtained from a strain of *Bacillus safensis* LAU 13. *Int. Nano Lett.,* **2015**, *5*(1), 29-35.
[http://dx.doi.org/10.1007/s40089-014-0133-4]

[21]   Mondal, A.H.; Yadav, D.; Mitra, S.; Mukhopadhyay, K. Biosynthesis of Silver Nanoparticles Using Culture Supernatant of *Shewanella* sp. ARY1 and Their Antibacterial Activity. *Int. J. Nanomedicine,* **2020**, *15*, 8295-8310.
[http://dx.doi.org/10.2147/IJN.S274535] [PMID: 33149577]

[22]   Wei, X.; Luo, M.; Li, W.; Yang, L.; Liang, X.; Xu, L.; Kong, P.; Liu, H. Synthesis of silver nanoparticles by solar irradiation of cell-free *Bacillus amyloliquefaciens* extracts and AgNO3. *Bioresour. Technol.,* **2012**, *103*(1), 273-278.
       [http://dx.doi.org/10.1016/j.biortech.2011.09.118] [PMID: 22019398]

[23]   Ramalingam, V.; Rajaram, R.; PremKumar, C.; Santhanam, P.; Dhinesh, P.; Vinothkumar, S.; Kaleshkumar, K. Biosynthesis of silver nanoparticles from deep sea bacterium Pseudomonas aeruginosa JQ989348 for antimicrobial, antibiofilm, and cytotoxic activity. *J. Basic Microbiol.,* **2014**, *54*(9), 928-936.
       [http://dx.doi.org/10.1002/jobm.201300514] [PMID: 24136453]

[24]   Saravanan, M.; Barik, S.K.; MubarakAli, D.; Prakash, P.; Pugazhendhi, A. Synthesis of silver nanoparticles from *Bacillus brevis* (NCIM 2533) and their antibacterial activity against pathogenic bacteria. *Microb. Pathog.,* **2018**, *116*, 221-226.
       [http://dx.doi.org/10.1016/j.micpath.2018.01.038] [PMID: 29407231]

[25]   Hossain, A.; Hong, X.; Ibrahim, E.; Li, B.; Sun, G.; Meng, Y.; Wang, Y.; An, Q. Green Synthesis of Silver Nanoparticles with Culture Supernatant of a Bacterium *Pseudomonas rhodesiae* and Their Antibacterial Activity against Soft Rot Pathogen *Dickeya dadantii. Molecules,* **2019**, *24*(12), 2303.
       [http://dx.doi.org/10.3390/molecules24122303] [PMID: 31234369]

[26]   Castro-Mayorga, J.L.; Martínez-Abad, A.; Fabra, M.J.; Olivera, C.; Reis, M.; Lagarón, J.M. Stabilization of antimicrobial silver nanoparticles by a polyhydroxyalkanoate obtained from mixed bacterial culture. *Int. J. Biol. Macromol.,* **2014**, *71*, 103-110.
       [http://dx.doi.org/10.1016/j.ijbiomac.2014.06.059] [PMID: 25043131]

[27]   Nadhe, S.B.; Singh, R.; Wadhwani, S.A.; Chopade, B.A. *Acinetobacter* sp. mediated synthesis of AgNPs, its optimization, characterization and synergistic antifungal activity against *C. albicans. J. Appl. Microbiol.,* **2019**, *127*(2), 445-458.
       [http://dx.doi.org/10.1111/jam.14305] [PMID: 31074075]

[28]   Singh, R.; Shedbalkar, U.U.; Nadhe, S.B.; Wadhwani, S.A.; Chopade, B.A. Lignin peroxidase mediated silver nanoparticle synthesis in *Acinetobacter* sp. *AMB Express,* **2017**, *7*(1), 226.
       [http://dx.doi.org/10.1186/s13568-017-0528-5] [PMID: 29273886]

[29]   Yang, J.; Wang, Q.; Wang, C.; Yang, R.; Ahmed, M.; Kumaran, S.; Velu, P.; Li, B. *Pseudomonas aeruginosa* synthesized silver nanoparticles inhibit cell proliferation and induce ROS mediated apoptosis in thyroid cancer cell line (TPC1). *Artif. Cells Nanomed. Biotechnol.,* **2020**, *48*(1), 800-809.
       [http://dx.doi.org/10.1080/21691401.2019.1687495] [PMID: 32432484]

[30]   Railean-Plugaru, V.; Pomastowski, P.; Wypij, M.; Szultka-Mlynska, M.; Rafinska, K.; Golinska, P.; Dahm, H.; Buszewski, B. Study of silver nanoparticles synthesized by acidophilic strain of Actinobacteria isolated from the of Picea sitchensis forest soil. *J. Appl. Microbiol.,* **2016**, *120*(5), 1250-1263.
       [http://dx.doi.org/10.1111/jam.13093] [PMID: 26864807]

[31]   Singh, R.; Wagh, P.; Wadhwani, S.; Gaidhani, S.; Kumbhar, A.; Bellare, J.; Chopade, B.A. Synthesis, optimization, and characterization of silver nanoparticles from *Acinetobacter calcoaceticus* and their enhanced antibacterial activity when combined with antibiotics. *Int. J. Nanomedicine,* **2013**, *8*, 4277-4290.
       [PMID: 24235826]

[32]   Gomaa, E.Z. Exopolysaccharide-mediated silver nanoparticles produced by Lactobacillus brevis NM101-1 as antibiotic adjuvant. *Microbiology,* **2016**, *85*(2), 207-219.
       [http://dx.doi.org/10.1134/S0026261716020077]

[33]   Oves, M.; Rauf, M.A.; Hussain, A.; Qari, H.A.; Khan, A.A.P.; Muhammad, P.; Rehman, M.T.; Alajmi, M.F.; Ismail, I.I.M. Antibacterial Silver Nanomaterial Synthesis From *Mesoflavibacter zeaxanthinifaciens* and Targeting Biofilm Formation. *Front. Pharmacol.,* **2019**, *10*, 801.
       [http://dx.doi.org/10.3389/fphar.2019.00801] [PMID: 31427961]

[34]   Thota, S.C.; Sreelatha, B. Biosynthesis of silver nanoparticles and exopolysaccharide using novel Thermophilic strain (Ts-1) of *Bacillus amyloliquefaciens. bioRxiv,* **2020.** [http://dx.doi.org/10.1101/2020.06.04.134742]

[35]   Garmasheva, I.; Kovalenko, N.; Voychuk, S.; Ostapchuk, A.; Livins'ka, O.; Oleschenko, L. *Lactobacillus* species mediated synthesis of silver nanoparticles and their antibacterial activity against opportunistic pathogens *in vitro. Bioimpacts,* **2016,** *6*(4), 219-223. [http://dx.doi.org/10.15171/bi.2016.29] [PMID: 28265538]

[36]   Zhang, J.; Yue, X.; Zeng, Y.; Hua, E.; Wang, M.; Sun, Y. Bacillus amyloliquefaciens levan and its silver nanoparticles with antimicrobial properties. *Biotechnol. Biotechnological Instrument,* **2018,** *32,* 1583-1589.

[37]   Elegbede, J.A.; Lateef, A.; Azeez, M.A.; Asafa, T.B.; Yekeen, T.A.; Oladipo, I.C.; Adebayo, E.A.; Beukes, L.S.; Gueguim-Kana, E.B. Fungal xylanases-mediated synthesis of silver nanoparticles for catalytic and biomedical applications. *IET Nanobiotechnol.,* **2018,** *12*(6), 857-863. [http://dx.doi.org/10.1049/iet-nbt.2017.0299] [PMID: 30104463]

[38]   Akther, T.; Vabeiryureilai Mathipi, ; Davoodbasha, M.; Srinivasan, H.; Srinivasan, H. Fungal-mediated synthesis of pharmaceutically active silver nanoparticles and anticancer property against A549 cells through apoptosis. *Environ. Sci. Pollut. Res. Int.,* **2019,** *26*(13), 13649-13657. [http://dx.doi.org/10.1007/s11356-019-04718-w] [PMID: 30919178]

[39]   Li, G.; He, D.; Qian, Y.; Guan, B.; Gao, S.; Cui, Y.; Yokoyama, K.; Wang, L. Fungus-mediated green synthesis of silver nanoparticles using *Aspergillus terreus. Int. J. Mol. Sci.,* **2012,** *13*(1), 466-476. [http://dx.doi.org/10.3390/ijms13010466] [PMID: 22312264]

[40]   Verma, V.C.; Kharwar, R.N.; Gange, A.C. Biosynthesis of antimicrobial silver nanoparticles by the endophytic fungus Aspergillus clavatus. *Nanomedicine (Lond.),* **2010,** *5*(1), 33-40. [http://dx.doi.org/10.2217/nnm.09.77] [PMID: 20025462]

[41]   Hulikere, M.; Joshi, C.G. Characterization, antioxidant and antimicrobial activity of silver nanoparticles synthesized using marine endophytic fungus- *Cladosporium cladosporioides. Process Biochem.,* **2019,** *82,* 199-204. [http://dx.doi.org/10.1016/j.procbio.2019.04.011]

[42]   Chen, X.; Yan, J-K.; Wu, J-Y. Characterization and antibacterial activity of silver nanoparticles prepared with a fungal exopolysaccharide in water. *Food Hydrocoll.,* **2016,** *53,* 69-74. [http://dx.doi.org/10.1016/j.foodhyd.2014.12.032]

[43]   Gudikandula, K.; Vadapally, P.; Charya, M.A.S. Biogenic synthesis of silver nanoparticles from white rot fungi: Their characterization and antibacterial studies. *Open Nano,* **2017,** *2,* 64-78. [http://dx.doi.org/10.1016/j.onano.2017.07.002]

[44]   Phanjom, P.; Ahmed, G. Effect of different physicochemical conditions on the synthesis of silver nanoparticles using fungal cell filtrate of Aspergillus oryzae (MTCC No. 1846) and their antibacterial effect. *Adv. Natural Sci: Nanosci. Nanotechnol,* **2017,** *8,* Article 045016.

[45]   Singh, T.; Jyoti, K.; Patnaik, A.; Singh, A.; Chauhan, R.; Chandel, S.S. Biosynthesis, characterization and antibacterial activity of silver nanoparticles using an endophytic fungal supernatant of *Raphanus sativus. J. Genet. Eng. Biotechnol.,* **2017,** *15*(1), 31-39. [http://dx.doi.org/10.1016/j.jgeb.2017.04.005] [PMID: 30647639]

[46]   Netala, V.R.; Kotakadi, V.S.; Bobbu, P.; Gaddam, S.A.; Tartte, V. Endophytic fungal isolate mediated biosynthesis of silver nanoparticles and their free radical scavenging activity and antimicrobial studies. *3 Biotech,* **2016,** *6,* Article 132.

[47]   Kobashigawa, J.M.; Robles, C.A.; Martínez Ricci, M.L.; Carmarán, C.C. Influence of strong bases on the synthesis of silver nanoparticles (AgNPs) using the ligninolytic fungi *Trametes trogii. Saudi J. Biol. Sci.,* **2019,** *26*(7), 1331-1337. [http://dx.doi.org/10.1016/j.sjbs.2018.09.006] [PMID: 31762592]

[48] Ameen, F.; Al-Homaidan, A.A.; Al-Sabri, A.; Almansob, A.; AlNAdhari, S. AlNAdhari, S. Antioxidant, anti-fungal and cytotoxic effects of silver nanoparticles synthesized using marine fungus *Cladosporium halotolerans. Appl. Nanosci.,* **2021.**
[http://dx.doi.org/10.1007/s13204-021-01874-9]

[49] Bagur, H.; Medidi, R.S.; Somu, P.; Choudhury, P.W.J.; Karua, C.S.; Guttala, P.K.; Melappa, G.; Poojari, C.C. Endophyte fungal isolate mediated biogenic synthesis and evaluation of biomedical applications of silver nanoparticles. *Mater. Technol.,* **2020,** 1-12.
[http://dx.doi.org/10.1080/10667857.2020.1819089]

[50] Chan, Y.S.; Mat Don, M. Biosynthesis and structural characterization of Ag nanoparticles from white rot fungi. *Mater. Sci. Eng. C,* **2013,** *33*(1), 282-288.
[http://dx.doi.org/10.1016/j.msec.2012.08.041] [PMID: 25428073]

[51] Sarsar, V.; Selwal, M.K.; Selwal, K.K. Biogenic synthesis, optimisation and antibacterial efficacy of extracellular silver nanoparticles using novel fungal isolate Aspergillus fumigatus MA. *IET Nanobiotechnol.,* **2016,** *10*(4), 215-221.
[http://dx.doi.org/10.1049/iet-nbt.2015.0058] [PMID: 27463792]

[52] Vahabi, K.; Mansoori, G.A.; Karimi, S. Biosynthesis of Silver Nanoparticles by Fungus *Trichoderma Reesei* (A Route for Large-Scale Production of AgNPs). *Insciences J.,* **2011,** *1*, 65-79.
[http://dx.doi.org/10.5640/insc.010165]

[53] Tyagi, S.; Tyagi, P.K.; Gola, D.; Chauhan, N.; Bharti, R.K. Extracellular synthesis of silver nanoparticles using entomopathogenic fungus: characterization and antibacterial potential. *SN Appl. Sci,* **2019,** *1*, Article 1545.

[54] Du, L.; Xu, Q.; Huang, M.; Xian, L.; Feng, J-X. Synthesis of small silver nanoparticles under light radiation by fungus Penicillium oxalicum and its application for the catalytic reduction of methylene blue. *Mater. Chem. Phys.,* **2015,** *160*, 40-47.
[http://dx.doi.org/10.1016/j.matchemphys.2015.04.003]

[55] Othman, A.M.; Elsayed, M.A.; Elshafei, A.M.; Hassan, M.M. Application of response surface methodology to optimize the extracellular fungal mediated nanosilver green synthesis. *J. Genet. Eng. Biotechnol.,* **2017,** *15*(2), 497-504.
[http://dx.doi.org/10.1016/j.jgeb.2017.08.003] [PMID: 30647692]

[56] Popli, D.; Anil, V.; Subramanyam, A.B.; M N, N.; v R, R.; Rao, S.N.; Rai, R.V.; Govindappa, M. Endophyte fungi, Cladosporium species-mediated synthesis of silver nanoparticles possessing in vitro antioxidant, anti-diabetic and anti-Alzheimer activity. *Artif. Cells Nanomed. Biotechnol.,* **2018,** *46*(sup1), 676-683.
[http://dx.doi.org/10.1080/21691401.2018.1434188] [PMID: 29400565]

[57] Salunkhe, R.B.; Patil, S.V.; Patil, C.D.; Salunke, B.K. Larvicidal potential of silver nanoparticles synthesized using fungus Cochliobolus lunatus against Aedes aegypti (Linnaeus, 1762) and Anopheles stephensi Liston (Diptera; Culicidae). *Parasitol. Res.,* **2011,** *109*(3), 823-831.
[http://dx.doi.org/10.1007/s00436-011-2328-1] [PMID: 21451993]

[58] Musarrat, J.; Dwivedi, S.; Singh, B.R.; Al-Khedhairy, A.A.; Azam, A.; Naqvi, A. Production of antimicrobial silver nanoparticles in water extracts of the fungus *Amylomyces rouxii* strain KSU-09. *Bioresour. Technol.,* **2010,** *101*(22), 8772-8776.
[http://dx.doi.org/10.1016/j.biortech.2010.06.065] [PMID: 20619641]

[59] Devi, L.S.; Joshi, S.R. Evaluation of the antimicrobial potency of silver nanoparticles biosynthesized by using an endophytic fungus, *Cryptosporiopsis ericae* PS4. *J. Microbiol.,* **2014,** *52*(8), 667-674.
[http://dx.doi.org/10.1007/s12275-014-4113-1] [PMID: 24994011]

[60] Chandankere, R.; Chelliah, J.; Subban, K.; Shanadrahalli, V.C.; Parvez, A.; Zabed, H.M.; Sharma, Y.C.; Qi, X. Pleiotropic Functions and Biological Potentials of Silver Nanoparticles Synthesized by an Endophytic Fungus. *Front. Bioeng. Biotechnol.,* **2020,** *8*, 95.
[http://dx.doi.org/10.3389/fbioe.2020.00095] [PMID: 32154230]

[61]   Neethu, S.; Midhun, S.J.; Radhakrishnan, E.K.; Jyothis, M. Green synthesized silver nanoparticles by marine endophytic fungus *Penicillium polonicum* and its antibacterial efficacy against biofilm forming, multidrug-resistant *Acinetobacter baumanii. Microb. Pathog.,* **2018**, *116*, 263-272. [http://dx.doi.org/10.1016/j.micpath.2018.01.033] [PMID: 29366864]

[62]   Nanda, A.; Nayak, B.K.; Krishnamoorthy, M. Antimicrobial properties of biogenic silver nanoparticles synthesized from phylloplane fungus, *Aspergillus tamarii. Biocatal. Agric. Biotechnol.,* **2018**, *16*, 225-228. [http://dx.doi.org/10.1016/j.bcab.2018.08.002]

[63]   Hu, X.; Saravanakumar, K.; Jin, T.; Wang, M-H. Mycosynthesis, characterization, anticancer and antibacterial activity of silver nanoparticles from endophytic fungus *Talaromyces purpureogenus. Int. J. Nanomedicine,* **2019**, *14*, 3427-3438. [http://dx.doi.org/10.2147/IJN.S200817] [PMID: 31190801]

[64]   Saxena, J.; Sharma, P.K.; Sharma, M.M.; Singh, A. Process optimization for green synthesis of silver nanoparticles by Sclerotinia sclerotiorum MTCC 8785 and evaluation of its antibacterial properties. *Springerplus,* **2016**, *5*(1), 861. [http://dx.doi.org/10.1186/s40064-016-2558-x] [PMID: 27386310]

[65]   Ammar, H.A.M.; El-Desouky, T.A. Green synthesis of nanosilver particles by *Aspergillus terreus* HA1N and *Penicillium expansum* HA2N and its antifungal activity against mycotoxigenic fungi. *J. Appl. Microbiol.,* **2016**, *121*(1), 89-100. [http://dx.doi.org/10.1111/jam.13140] [PMID: 27002915]

[66]   Jalal, M.; Ansari, M.A.; Alzohairy, M.A.; Ali, S.G.; Khan, H.M.; Almatroudi, A.; Raees, K. Biosynthesis of Silver Nanoparticles from Oropharyngeal *Candida glabrata* Isolates and Their Antimicrobial Activity against Clinical Strains of Bacteria and Fungi. *Nanomater. (MDPI),* **2018**, *8*, Article 586.

[67]   Fathima, B.S.; Balakrishnan, R.M. Biosynthesis and optimization of silver nanoparticles by endophytic fungus *Fusarium solani. Mater. Lett.,* **2014**, *132*, 428-431. [http://dx.doi.org/10.1016/j.matlet.2014.06.143]

[68]   Moustafa, M.T. Removal of pathogenic bacteria from wastewater using silver nanoparticles synthesized by two fungal species. *Water Sci,* **2017**, *31*(2), 164-176. [http://dx.doi.org/10.1016/j.wsj.2017.11.001]

[69]   Dar, M.A.; Ingle, A.; Rai, M. Enhanced antimicrobial activity of silver nanoparticles synthesized by *Cryphonectria* sp. evaluated singly and in combination with antibiotics. *Nanomedicine,* **2013**, *9*(1), 105-110. [http://dx.doi.org/10.1016/j.nano.2012.04.007] [PMID: 22633901]

[70]   Singh, D.; Rathod, V.; Ninganagouda, S.; Hiremath, J.; Singh, A.K.; Mathew, J. Optimization and Characterization of Silver Nanoparticle by Endophytic Fungi *Penicillium* sp. Isolated from *Curcuma longa* (Turmeric) and Application Studies against MDR E. coli and S. aureus. *Bioinorg. Chem. Appl.,* **2014**, *2014*: 408021. [http://dx.doi.org/10.1155/2014/408021] [PMID: 24639625]

[71]   Hamedi, S.; Shojaosadati, S.A.; Shokrollahzadeh, S.; Hashemi-Najafabadi, S. Extracellular biosynthesis of silver nanoparticles using a novel and non-pathogenic fungus, *Neurospora intermedia*: controlled synthesis and antibacterial activity. *World J. Microbiol. Biotechnol.,* **2014**, *30*(2), 693-704. [http://dx.doi.org/10.1007/s11274-013-1417-y] [PMID: 24068530]

[72]   Yin, Y.; Yang, X.; Hu, L.; Tan, Z.; Zhao, L.; Zhang, Z.; Liu, J.; Jiang, G. Superoxide-Mediated Extracellular Biosynthesis of Silver Nanoparticles by the Fungus *Fusarium oxysporum. Environ. Sci. Technol. Lett.,* **2016**, *3*(4), 160-165. [http://dx.doi.org/10.1021/acs.estlett.6b00066]

[73]   Barbosa, A.C.M.S.; Silva, L.P.C.; Ferraz, C.M.; Tobias, F.L.; de Araújo, J.V.; Loureiro, B.; Braga, G.M.A.M.; Veloso, F.B.R.; Soares, F.E.F.; Fronza, M.; Braga, F.R. Nematicidal activity of silver

nanoparticles from the fungus *Duddingtonia flagrans. Int. J. Nanomedicine,* **2019**, *14*, 2341-2348.
[http://dx.doi.org/10.2147/IJN.S193679] [PMID: 31040660]

[74] Prasad, T.N.V.K.V.; Kambala, V.S.R.; Naidu, R. Phyconanotechnology: synthesis of silver nanoparticles using brown marine algae *Cystophora moniliformis* and their characterization. *J. Appl. Phycol.,* **2013**, *25*(1), 177-182.
[http://dx.doi.org/10.1007/s10811-012-9851-z]

[75] Aboelfetoh, E.F.; El-Shenody, R.A.; Ghobara, M.M. Eco-friendly synthesis of silver nanoparticles using green algae (*Caulerpa serrulata*): reaction optimization, catalytic and antibacterial activities. *Environ. Monit. Assess.,* **2017**, *189*(7), 349.
[http://dx.doi.org/10.1007/s10661-017-6033-0] [PMID: 28646435]

[76] Kathiraven, T.; Sundaramanickam, A.; Shanmugam, N.; Balasubramanian, T. Green synthesis of silver nanoparticles using marine algae *Caulerpa racemosa* and their antibacterial activity against some human pathogens. *Appl. Nanosci.,* **2015**, *5*(4), 499-504.
[http://dx.doi.org/10.1007/s13204-014-0341-2]

[77] Khalifa, K.S.; Hamouda, R.A.; Hamza, D.H.A. *In vitro* antitumor activity of silver nanoparticles biosynthesized by marine algae. *Dig. J. Nanomater. Biostruct.,* **2016**, *11*, 213-221.

[78] Venkatesan, J.; Kim, S-K; Shim, M.S. Antimicrobial, Antioxidant, and Anticancer Activities of Biosynthesized Silver Nanoparticles Using Marine Algae *Ecklonia cava. Nanomater (MDPI),* **2016**, *6*, Article 235.

[79] Raouf, N.A.; Alharbi, R.M.; Al-Enazi, N.M.; Alkhulaifi, M.M.; Ibraheem, I.B.M. Rapid biosynthesis of silver nanoparticles using the marine red alga Laurencia catarinensis and their characterization. *Beni. Suef Univ. J. Basic Appl. Sci.,* **2018**, *7*(1), 150-157.
[http://dx.doi.org/10.1016/j.bjbas.2017.10.003]

[80] Moshfegh, A.; Jalali, A.; Salehzadeh, A.; Jozani, A.S. Biological synthesis of silver nanoparticles by cell-free extract of Polysiphonia algae and their anticancer activity against breast cancer MCF-7 cell lines. *Micro & Nano Lett.,* **2019**, *14*(5), 581-584.
[http://dx.doi.org/10.1049/mnl.2018.5260]

[81] Sinha, S.N.; Paul, D.; Halder, N.; Sengupta, D.; Patra, S.K. Green synthesis of silver nanoparticles using fresh water green alga *Pithophora oedogonia* (Mont.) Wittrock and evaluation of their antibacterial activity. *Appl. Nanosci.,* **2015**, *5*(6), 703-709.
[http://dx.doi.org/10.1007/s13204-014-0366-6]

[82] Massironi, A.; Morelli, A.; Grassi, L.; Puppi, D.; Braccini, S.; Maisetta, G.; Esin, S.; Batoni, G.; Della Pina, C.; Chiellini, F. Ulvan as novel reducing and stabilizing agent from renewable algal biomass: Application to green synthesis of silver nanoparticles. *Carbohydr. Polym.,* **2019**, *203*, 310-321.
[http://dx.doi.org/10.1016/j.carbpol.2018.09.066] [PMID: 30318218]

[83] El-Rafie, H.M.; El-Rafie, M.H.; Zahran, M.K. Green synthesis of silver nanoparticles using polysaccharides extracted from marine macro algae. *Carbohydr. Polym.,* **2013**, *96*(2), 403-410.
[http://dx.doi.org/10.1016/j.carbpol.2013.03.071] [PMID: 23768580]

[84] Edison, T.N.J.I.; Atchudan, R.; Kamal, C.; Lee, Y.R. *Caulerpa racemosa*: a marine green alga for eco-friendly synthesis of silver nanoparticles and its catalytic degradation of methylene blue. *Bioprocess Biosyst. Eng.,* **2016**, *39*(9), 1401-1408.
[http://dx.doi.org/10.1007/s00449-016-1616-7] [PMID: 27129459]

[85] Ozturk, B.Y.; Gursu, B.Y.; Dag, I. Antibiofilm and antimicrobial activities of green synthesized silver nanoparticles using marine red algae *Gelidium corneum. Process Biochem.,* **2020**, *89*, 208-219.
[http://dx.doi.org/10.1016/j.procbio.2019.10.027]

[86] Gopu, M.; Kumar, P.; Selvankumar, T.; Senthilkumar, B.; Sudhakar, C.; Govarthanan, M.; Selva Kumar, R.; Selvam, K. Green biomimetic silver nanoparticles utilizing the red algae *Amphiroa rigida* and its potent antibacterial, cytotoxicity and larvicidal efficiency. *Bioprocess Biosyst. Eng.,* **2021**, *44*(2), 217-223.

[http://dx.doi.org/10.1007/s00449-020-02426-1] [PMID: 32803487]

[87]   Borah, D.; Das, N.; Das, N.; Bhattacharjee, A.; Sarmah, P.; Ghosh, K.; Chandel, M.; Rout, J.; Pandey, P.; Ghosh, N.N.; Bhattacharjee, C.R. Alga-mediated facile green synthesis of silver nanoparticles: Photophysical, catalytic and antibacterial activity. *Appl. Organomet. Chem.,* **2020**, *34*(5): e5597.
[http://dx.doi.org/10.1002/aoc.5597]

[88]   Priyadharshini, R.I.; Prasannaraj, G.; Geetha, N.; Venkatachalam, P. Microwave-mediated extracellular synthesis of metallic silver and zinc oxide nanoparticles using macro-algae (*Gracilaria edulis*) extracts and its anticancer activity against human PC3 cell lines. *Appl. Biochem. Biotechnol.,* **2014**, *174*(8), 2777-2790.
[http://dx.doi.org/10.1007/s12010-014-1225-3] [PMID: 25380639]

[89]   Abdel-Raouf, N.; Al-Enazi, N.M.; Ibraheem, I.B.M.; Alharbi, R.M.; Alkhulaifi, M.M. Biosynthesis of silver nanoparticles by using of the marine brown alga *Padina pavonia* and their characterization. *Saudi J. Biol. Sci.,* **2019**, *26*(6), 1207-1215.
[http://dx.doi.org/10.1016/j.sjbs.2018.01.007] [PMID: 31516350]

[90]   Ibraheem, I.B.M.; Abd Elaziz, B.E.E.; Saad, W.F.; Fathy, W.A. Green Biosynthesis of Silver Nanoparticles Using Marine Red Algae *Acanthophora specifera* and its Antimicrobial Activity. *J. Nanomed. Nanotechnol.,* **2016**, *7*: 1000409.

[91]   Elgamouz, A.; Idriss, H.; Nassab, C.; Bihi, A.; Bajou, K.; Hasan, K.; Haija, M.A.; Patole, S.P. Green Synthesis, Characterization, Antimicrobial, Anti-Cancer, and Optimization of Colorimetric Sensing of Hydrogen Peroxide of Algae Extract Capped Silver Nanoparticles. *Nanomater. (MDPI),* **2020**, *10*, Article 1861.

[92]   Xu, Y.; Hou, Y.; Wang, Y.; Wang, Y.; Li, T.; Song, C.; Wei, N.; Wang, Q. Sensitive and selective detection of Cu^{2+} ions based on fluorescent Ag nanoparticles synthesized by R-phycoerythrin from marine algae Porphyra yezoensis. *Ecotoxicol. Environ. Saf.,* **2019**, *168*, 356-362.
[http://dx.doi.org/10.1016/j.ecoenv.2018.10.102] [PMID: 30391840]

[93]   Vieira, A.P.; Stein, E.M.; Andreguetti, D.X.; Colepicolo, P.; Ferreira, C.A.M. reparation of silver nanoparticles using aqueous extracts of the red algae *Laurencia aldingensis* and *Laurenciella* sp. and their cytotoxic activities. *J. Appl. Phycol.,* **2016**, *28*(4), 2615-2622.
[http://dx.doi.org/10.1007/s10811-015-0757-4]

[94]   Muthusamy, G.; Thangasamy, S.; Raja, M.; Chinnappan, S.; Kandasamy, S. Biosynthesis of silver nanoparticles from Spirulina microalgae and its antibacterial activity. *Environ. Sci. Pollut. Res. Int.,* **2017**, *24*(23), 19459-19464.
[http://dx.doi.org/10.1007/s11356-017-9772-0] [PMID: 28730357]

[95]   Velgosova, O.; Dolinska, S.; Veselovsky, L. Possibilities of modification of green algae P. kessleri extracts composition and their influence on silver nanoparticles synthesis. *Mol. Cryst. Liq. Cryst. (Phila. Pa.),* **2020**, *711*(1), 41-49.
[http://dx.doi.org/10.1080/15421406.2020.1840689]

[96]   Bao, Z.; Cao, J.; Kang, G.; Lan, C.Q. Effects of reaction conditions on light-dependent silver nanoparticle biosynthesis mediated by cell extract of green alga Neochloris oleoabundans. *Environ. Sci. Pollut. Res. Int.,* **2019**, *26*(3), 2873-2881.
[http://dx.doi.org/10.1007/s11356-018-3843-8] [PMID: 30499085]

[97]   Yan, N.; Wang, W-X. Novel Imaging of Silver Nanoparticle Uptake by a Unicellular Alga and Trophic Transfer to *Daphnia magna. Environ. Sci. Technol.,* **2021**, *55*(8), 5143-5151.
[http://dx.doi.org/10.1021/acs.est.0c08588] [PMID: 33726495]

[98]   Kalimuthu, K.; Paneelselvam, C.; Chou, C.; Lin, S-M.; Tseng, L-C.; Tsai, K-H.; Murugan, K.; Hwang, J-S. Predatory efficiency of the copepod *Megacyclops formosanus* and toxic effect of the red alga Gracilaria firma-synthesized silver nanoparticles against the dengue vector *Aedes aegypti. Hydrobiologia,* **2017**, *785*(1), 359-372.
[http://dx.doi.org/10.1007/s10750-016-2943-z]

[99] Aragao, A.P.; Oliveira, T.M.; Quelemes, P.V.; Luana, M.; Perfeito, G.; Araujo, M.C.; Santiago, J.A.S.; Cardoso, V.S.; Quaresma, P.; Leite, J.R.S.A.; Silva, D.A. Green synthesis of silver nanoparticles using the seaweed *Gracilaria birdiae* and their antibacterial activity. *Arab. J. Chem.,* **2019**, *12*(8), 4182-4188.
      [http://dx.doi.org/10.1016/j.arabjc.2016.04.014]

[100] Rajkumar, R.; Ezhumalai, G.; Gnanadesign, M. A green approach for the synthesis of silver nanoparticles by Chlorella vulgaris and its application in photocatalytic dye degradation activity. *Environ. Technol. Innovat.,* **2021**, *21*, Article 101282.

[101] Selvaraj, P.; Neethu, E.; Rathika, P.; Jayaseeli, J.P.R.; Jermy, B.R.; Azeez, S.A.; Borgio, J.F.; Dhas, T.S. Antibacterial potentials of methanolic extract and silver nanoparticles from marine algae. *Biocatal. Agric. Biotechnol.,* **2020**, *28*: 101719.
      [http://dx.doi.org/10.1016/j.bcab.2020.101719]

[102] Bhuyar, P.; Rahim, M.H.A.; Sundaraju, S.; Ramaraj, R.; Maniam, G.P.; Govindan, N. Synthesis of silver nanoparticles using marine macroalgae Padina sp. and its antibacterial activity towards pathogenic bacteria. *Beni. Suef Univ. J. Basic Appl. Sci.,* **2020**, *9*(1), 3.
      [http://dx.doi.org/10.1186/s43088-019-0031-y]

[103] Acharya, D.; Satapathy, S.; Thathapudi, J.J.; Somu, P.; Mishra, G. Biogenic synthesis of silver nanoparticles using marine algae *Cladophora glomerata* and evaluation of apoptotic effects in human colon cancer cells. *Mater. Technol.,* **2020**, 1-12.
      [http://dx.doi.org/10.1080/10667857.2020.1863597]

[104] Satishkumar, R.S.; Sundaramanickam, A.; Srinath, R.; Ramesh, T.; Saranya, K.; Meena, M.; Surya, P. Green synthesis of silver nanoparticles by bloom forming marine microalgae *Trichodesmium erythraeum* and its applications in antioxidant, drug-resistant bacteria, and cytotoxicity activity. *J. Saudi Chem. Soc.,* **2019**, *23*(8), 1180-1191.
      [http://dx.doi.org/10.1016/j.jscs.2019.07.008]

[105] Daglioglu, Y.; Ozturk, B.Y. A novel intracellular synthesis of silver nanoparticles using *Desmodesmus* sp. (Scenedesmaceae): different methods of pigment change. *Rend. Lincei Sci. Fis. Nat.,* **2019**, *30*(3), 611-621.
      [http://dx.doi.org/10.1007/s12210-019-00822-8]

[106] Roseline, T.A.; Murugan, M.; Sudhakar, M.P.; Arunkumar, K. ArunKumar, K. Nanopesticidal potential of silver nanocomposites synthesized from the aqueous extracts of red seaweeds. *Environ. Technol. Innovat.,* **2019**, *13*, 82-93.
      [http://dx.doi.org/10.1016/j.eti.2018.10.005]

[107] Kannan, R.R.R.; Stirk, W.A.; Staden, J.V. Synthesis of silver nanoparticles using the seaweed *Codium capitatum* P.C. *Silva (Chlorophyceae),* **2013**, *86*, 1-4.

[108] Rahman, A.; Kumar, S.; Bafana, A.; Lin, J.; Dahoumane, S.A.; Jeffryes, C. A Mechanistic View of the Light-Induced Synthesis of Silver Nanoparticles Using Extracellular Polymeric Substances of *Chlamydomonas reinhardtii. Molecules,* **2019**, *24*(19), 3506.
      [http://dx.doi.org/10.3390/molecules24193506] [PMID: 31569641]

[109] Arya, A.; Gupta, K.; Chundawat, T.S.; Vaya, D. Biogenic Synthesis of Copper and Silver Nanoparticles Using Green Alga *Botryococcus braunii* and Its Antimicrobial Activity. *Bioinorg. Chem. Appl.,* **2018**, *2018*: 7879403.
      [http://dx.doi.org/10.1155/2018/7879403] [PMID: 30420873]

[110] Kim, D-Y.; Saratale, R.G.; Shinde, S.; Syed, A.; Ameen, F.; Ghodake, G. Green synthesis of silver nanoparticles using *Laminaria japonica* extract: Characterization and seedling growth assessment. *J. Clean. Prod.,* **2018**, *172*, 2910-2918.
      [http://dx.doi.org/10.1016/j.jclepro.2017.11.123]

[111] Kumar, R.; Roopan, S.M.; Prabhakarn, A.; Khanna, V.G.; Chakroborty, S. Agricultural waste Annona squamosa peel extract: biosynthesis of silver nanoparticles. *Spectrochim. Acta A Mol. Biomol.*

Spectrosc., **2012**, *90*, 173-176.
[http://dx.doi.org/10.1016/j.saa.2012.01.029] [PMID: 22336049]

[112] Sinsinwar, S.; Sarkar, M.K.; Suriya, K.R.; Nithyanand, P.; Vadivel, V. Use of agricultural waste (coconut shell) for the synthesis of silver nanoparticles and evaluation of their antibacterial activity against selected human pathogens. *Microb. Pathog.,* **2018**, *124*, 30-37.
[http://dx.doi.org/10.1016/j.micpath.2018.08.025] [PMID: 30120992]

[113] Gul, A.R.; Shaheen, F.; Rafique, R.; Bal, J.; Waseem, S.; Park, T.J. Grass-mediated biogenic synthesis of silver nanoparticles and their drug delivery evaluation: A biocompatible anti-cancer therapy. *Chem. Eng. J.,* **2021**, *407*: 127202.
[http://dx.doi.org/10.1016/j.cej.2020.127202]

[114] Omran, B.A.; Nassar, H.N.; Fatthallah, N.A.; Hamdy, A.; El-Shatoury, E.H.; El-Gendy, N.S. Waste upcycling of *Citrus sinensis* peels as a green route for the synthesis of silver nanoparticles. *Energy Sources A Recovery Util. Environ. Effects,* **2018**, *40*(2), 227-236.
[http://dx.doi.org/10.1080/15567036.2017.1410597]

[115] Das, G.; Shin, H-S.; Kumar, A.; Vishnuprasad, C.N.; Patra, J.K. Photo-mediated optimized synthesis of silver nanoparticles using the extracts of outer shell fibre of *Cocos nucifera* L. fruit and detection of its antioxidant, cytotoxicity and antibacterial potential. *Saudi J. Biol. Sci.,* **2021**, *28*(1), 980-987.
[http://dx.doi.org/10.1016/j.sjbs.2020.11.022] [PMID: 33424390]

[116] Devadiga, A.; Shetty, K.V.; Saidutta, M.B. Timber industry waste-teak (Tectona grandis Linn.) leaf extract mediated synthesis of antibacterial silver nanoparticles. *Int. Nano Lett.,* **2015**, *5*(4), 205-214.
[http://dx.doi.org/10.1007/s40089-015-0157-4]

[117] Das, G.; Patra, J.K.; Debnath, T.; Ansari, A.; Shin, H.S. Investigation of antioxidant, antibacterial, antidiabetic, and cytotoxicity potential of silver nanoparticles synthesized using the outer peel extract of Ananas comosus (L.). *PLoS One,* **2019**, *14*(8): e0220950.
[http://dx.doi.org/10.1371/journal.pone.0220950] [PMID: 31404086]

[118] Lateef, A.; Azeez, M.A.; Asafa, T.B.; Yekeen, T.A.; Akinboro, A.; Oladipo, I.C.; Azeez, L.; Ojo, S.A.; Kana, E.B.G.; Beukes, L.S. Cocoa pod husk extract-mediated biosynthesis of silver nanoparticles: its antimicrobial, antioxidant and larvicidal activities. *J. Nanostructure Chem.,* **2016**, *6*(2), 159-169.
[http://dx.doi.org/10.1007/s40097-016-0191-4]

[119] Carbone, K.; Paliotta, M.; Micheli, L.; Mazzuca, C.; Cacciotti, I.; Nocente, F.; Ciampa, A.; Abate, M.T.D. A completely green approach to the synthesis of dendritic silver nanostructures starting from white grape pomace as a potential nanofactory. *Arab. J. Chem.,* **2019**, *12*(5), 597-609.
[http://dx.doi.org/10.1016/j.arabjc.2018.08.001]

[120] Das, G.; Shin, H.S.; Patra, J.K. Comparative Assessment of Antioxidant, Anti-Diabetic and Cytotoxic Effects of Three Peel/Shell Food Waste Extract-Mediated Silver Nanoparticles. *Int. J. Nanomedicine,* **2020**, *15*, 9075-9088.
[http://dx.doi.org/10.2147/IJN.S277625] [PMID: 33235452]

[121] Perera, K.M.K.G.; Kuruppu, K.A.S.S.; Chamara, A.M.R.; Thiripuranathar, G. Characterization of spherical Ag nanoparticles synthesized from the agricultural wastes of *Garcinia mangostana* and *Nephelium lappaceum* and their applications as a photo catalyzer and fluorescence quencher. *SN Appl. Sci.,* **2020**, *2*, Article 1974.

[122] Ndikau, M.; Noah, N.M.; Andala, D.M.; Masika, E. Green Synthesis and Characterization of Silver Nanoparticles Using *Citrullus lanatus* Fruit Rind Extract. *Int. J. Anal. Chem.,* **2017**, *2017*: 8108504.
[http://dx.doi.org/10.1155/2017/8108504] [PMID: 28316627]

[123] Kumar, B.; Smita, K.; Cumbal, L.; Debut, A. Sacha inchi (*Plukenetia volubilis* L.) shell biomass for synthesis of silver nanocatalyst. *J. Saudi Chem. Soc.,* **2017**, *21*, 293-298.
[http://dx.doi.org/10.1016/j.jscs.2014.03.005]

[124] Kokila, T.; Ramesh, P.S.; Geetha, D. Biosynthesis of AgNPs using Carica Papaya peel extract and evaluation of its antioxidant and antimicrobial activities. *Ecotoxicol. Environ. Saf.,* **2016**, *134*(Pt 2),

467-473.
[http://dx.doi.org/10.1016/j.ecoenv.2016.03.021] [PMID: 27156649]

[125]   Omran, B.A.; Aboelazayem, O.; Nassar, H.N.; El-Salamony, R.A.; El-Gendy, N.S. Biovalorization of mandarin waste peels into silver nanoparticles and activated carbon. *Int. J. Environ. Sci. Technol.,* **2021,** *18*(5), 1119-1134.
[http://dx.doi.org/10.1007/s13762-020-02873-z]

[126]   Cheng, T-H.; Yang, Z-Y.; Tang, R-C.; Zhai, A-D. Functionalization of silk by silver nanoparticles synthesized using the aqueous extract from tea stem waste. *J. Mater. Res. Technol.,* **2020,** *9*(3), 4538-4549.
[http://dx.doi.org/10.1016/j.jmrt.2020.02.081]

[127]   Patra, J.K.; Das, G.; Kumar, A.; Ansari, A.; Kim, H.; Shin, H-S. Photo-mediated Biosynthesis of Silver Nanoparticles Using the Non-edible Accrescent Fruiting Calyx of *Physalis peruviana* L. Fruits and Investigation of its Radical Scavenging Potential and Cytotoxicity Activities. *J. Photochem. Photobiol. B,* **2018,** *188*, 116-125.
[http://dx.doi.org/10.1016/j.jphotobiol.2018.08.004] [PMID: 30266015]

[128]   Song, C.; Ye, F.; Liu, S.; Li, F.; Huang, Y.; Ji, R.; Zhao, L. Thorough utilization of rice husk: metabolite extracts for silver nanocomposite biosynthesis and residues for silica nanomaterials fabrication. *New J. Chem.,* **2019,** *43*(23), 9201-9209.
[http://dx.doi.org/10.1039/C9NJ01926J]

[129]   Srikhao, N.; Kasemsiri, P.; Lorwanishpaisarn, N.; Okhawilai, M. Green synthesis of silver nanoparticles using sugarcane leaves extract for colorimetric detection of ammonia and hydrogen peroxide. *Res. Chem. Intermed.,* **2021,** *47*(3), 1269-1283.
[http://dx.doi.org/10.1007/s11164-020-04354-x]

[130]   Yao, P.; Zhang, J.; Xing, T.; Chen, G.; Tao, R.; Choo, K-H. Green synthesis of silver nanoparticles using grape seed extract and their application for reductive catalysis of Direct Orange 26. *J. Ind. Eng. Chem.,* **2018,** *58*, 74-79.
[http://dx.doi.org/10.1016/j.jiec.2017.09.009]

[131]   Islam, S.; Butola, B.S.; Gupta, A.; Roy, A. Multifunctional finishing of cellulosic fabric *via* facile, rapid *in-situ* green synthesis of AgNPs using pomegranate peel extract biomolecules. *Sustain. Chem. Pharm.,* **2019,** *12*: 100135.
[http://dx.doi.org/10.1016/j.scp.2019.100135]

[132]   Harish, B.S.; Uppuluri, K.B.; Anbazhagan, V. Synthesis of fibrinolytic active silver nanoparticle using wheat bran xylan as a reducing and stabilizing agent. *Carbohydr. Polym.,* **2015,** *132*, 104-110.
[http://dx.doi.org/10.1016/j.carbpol.2015.06.069] [PMID: 26256330]

[133]   Lateef, A.; Azeez, M.A.; Asafa, T.B.; Yekeen, T.A.; Akinboro, A.; Oladipo, I.C.; Azeez, L.; Ajibade, S.E.; Ojo, S.A.; Kana, E.B.G.; Beukes, L.S. Biogenic synthesis of silver nanoparticles using a pod extract of Cola nitida: Antibacterial and antioxidant activities and application as a paint additive. *J. Taibah Univ. Sci.,* **2016,** *10*(4), 551-562.
[http://dx.doi.org/10.1016/j.jtusci.2015.10.010]

[134]   Devanesan, S.; AlSalhi, M.S.; Balaji, R.V.; Ranjitsingh, A.J.A.; Ahamed, A.; Alfuraydi, A.A.; AlQahtani, F.Y.; Aleanizy, F.S.; Othman, A.H. Antimicrobial and Cytotoxicity Effects of Synthesized Silver Nanoparticles from Punica granatum Peel Extract. *Nanoscale Res. Lett.,* **2018,** *13*(1), 315.
[http://dx.doi.org/10.1186/s11671-018-2731-y] [PMID: 30288618]

[135]   Bordbar, M. Biosynthesis of Ag/almond shell nanocomposite as a cost-effective and efficient catalyst for degradation of 4-nitrophenol and organic dyes. *RSC Advances,* **2017,** *7*(1), 180-189.
[http://dx.doi.org/10.1039/C6RA24977A]

[136]   Pani, A.; Lee, J.H.; Yun, S.I. Autoclave mediated one-pot-one-minute synthesis of AgNPs and Au-Ag nanocomposite from *Melia azedarach* bark extract with antimicrobial activity against food pathogens. *Chem. Cent. J.,* **2016,** *10*(1), 15.

[http://dx.doi.org/10.1186/s13065-016-0157-0] [PMID: 27042205]

[137]   Chutrakulwong, F.; Thamaphat, K.; Limsuwan, P. Photo-irradiation induced green synthesis of highly stable silver nanoparticles using durian rind biomass: effects of light intensity, exposure time and pH on silver nanoparticles formation. *J. Phys. Commun.,* **2020**, *4*, Article 095015.

[138]   Kumar, B.; Kumari, S.; Cumbal, L.; Debut, A. *Lantana camara* berry for the synthesis of silver nanoparticles. *Asian Pac. J. Trop. Biomed.,* **2015**, *5*(3), 192-195.
[http://dx.doi.org/10.1016/S2221-1691(15)30005-8]

[139]   Bagherzade, G.; Tavakoli, M.M.; Namaei, M.H. Green synthesis of silver nanoparticles using aqueous extract of saffron (Crocus sativus L.) wastages and its antibacterial activity against six bacteria. *Asian Pac. J. Trop. Biomed.,* **2017**, *7*(3), 227-233.
[http://dx.doi.org/10.1016/j.apjtb.2016.12.014]

[140]   Kokila, T.; Ramesh, P.S.; Geetha, D. Biosynthesis of silver nanoparticles from Cavendish banana peel extract and its antibacterial and free radical scavenging assay: a novel biological approach. *Appl. Nanosci.,* **2015**, *5*(8), 911-920.
[http://dx.doi.org/10.1007/s13204-015-0401-2]

[141]   Skiba, M.I.; Vorobyova, V.I. Synthesis of Silver Nanoparticles Using Orange Peel Extract Prepared by Plasmochemical Extraction Method and Degradation of Methylene Blue under Solar Irradiation. *Adv. Mater. Sci. Eng.,* **2019**, *2019*: 8306015.
[http://dx.doi.org/10.1155/2019/8306015]

[142]   Velu, M.; Lee, J-H; Chang, W-S; Lovanh, N.; Park, Y-J; Jayanthi, P.; Palanivel, V.; Oh, B-T. Fabrication, optimization, and characterization of noble silver nanoparticles from sugarcane leaf (*Saccharum officinarum*) extract for antifungal application. *3 Biotech.,* **2017**, *7*, Article 147.

[143]   Govarthanan, M.; Seo, Y-S.; Lee, K-J.; Jung, I-B.; Ju, H-J.; Kim, J.S.; Cho, M.; Kamala-Kannan, S.; Oh, B.T. Low-cost and eco-friendly synthesis of silver nanoparticles using coconut (*Cocos nucifera*) oil cake extract and its antibacterial activity. *Artif. Cells Nanomed. Biotechnol.,* **2016**, *44*(8), 1878-1882.
[http://dx.doi.org/10.3109/21691401.2015.1111230] [PMID: 26855063]

[144]   Jiang, Q.; Luo, B.; Wu, Z.; Gu, B.; Xu, C.; Li, X.; Wang, X. Corn stalk/AgNPs modified chitin composite hemostatic sponge with high absorbency, rapid shape recovery and promoting wound healing ability. *Chem. Eng. J.,* **2021**, *421*: 129815.
[http://dx.doi.org/10.1016/j.cej.2021.129815]

[145]   Liu, Y-S.; Chang, Y-C.; Chen, H-H. Silver nanoparticle biosynthesis by using phenolic acids in rice husk extract as reducing agents and dispersants. *J. Food Drug Anal.,* **2018**, *26*(2), 649-656.
[http://dx.doi.org/10.1016/j.jfda.2017.07.005] [PMID: 29567234]

[146]   He, D.; Ikeda-Ohno, A.; Boland, D.D.; Waite, T.D. Synthesis and characterization of antibacterial silver nanoparticle-impregnated rice husks and rice husk ash. *Environ. Sci. Technol.,* **2013**, *47*(10), 5276-5284.
[http://dx.doi.org/10.1021/es303890y] [PMID: 23614704]

[147]   Lieu, Y-S.; Chang, Y-C.; Chen, H-H. Synthesis of silver nanoparticles by using rice husk extracts prepared with acid–alkali pretreatment extraction process. *J. Cereal Sci.,* **2018**, *82*, 106-112.
[http://dx.doi.org/10.1016/j.jcs.2018.06.002]

[148]   Azam, F.A.A.; Shamsudin, R.; Ng, M.H.; Ahmad, A.; Afiq, M.; Akbar, M.; Rashidbenam, Z. Silver-doped pseudowollastonite synthesized from rice husk ash: Antimicrobial evaluation, bioactivity and cytotoxic effects on human mesenchymal stem cells. *Ceram. Int.,* **2018**, *44*(10), 11381-11389.
[http://dx.doi.org/10.1016/j.ceramint.2018.03.189]

CHAPTER 3

Surface Functionalization of AgNPs

Abstract: The surface fabrication of AgNPs with different functionalities provides supplementary appendages for extended applications. These functionalities provide base for further anchoring, conjugation, or interacting with biomolecules, polymers, and synthetic chemical molecules deliberated at their biological target. The surface functionalities mask the engineered surface and surface properties of AgNPs and prevent them from modifications *in situ*. Moreover, the surface functionalized AgNPs with functional head groups enables their further conjugation to the nucleic acids, aptamers, peptides, and lipopolysaccharides that provide excellent applications in molecular medicine. This chapter presents the surface functionalization of AgNPs with natural molecules such as polysaccharides, amino acids, nucleic acids, peptides and their applications.

Keywords: Biotolerance, Conjugation, Molecular medicine, Surface functionalization, Toxicity.

1. INTRODUCTION

The physiological and environmental toxicity associated with the AgNPs limits their ubiquitous applications. The physiological toxicity of AgNPs mainly arises due to the production of Ag^+ ions, which trigger the cellular oxidative stress by generating reactive oxygen species (ROS). The ROS adversely affect the functioning of cell organelles, and disrupt the biomacromolecules resulting in anomalous cellular functioning. Similarly, the direct biological exposure of AgNPs results in the development of proteinaceous surface corona, which conceals the original engineered surface of the nanoparticles. This event results in erroneous protein expression causing functional deformities in cells. The environmental toxicity and the hazardous implications towards agroecosystems mainly arise due to complex physicochemical transformations of AgNPs, including aggregation, oxidative dissolution, sulfidation, chlorination, and photochemical transformations, in addition to a direct exposure towards the biosphere, atmosphere and lithosphere. Surface functionalization with a suitable ligand or moiety mitigates the toxicity of AgNPs, and provides additional functionalities for extended applications *via* chemical or physical conjugation to other molecules. As such, the linking of AgNPs with biomolecules offers extensive uses in molecular medicine, enhancing the bioavailability of conjugant drug molecule, and presenting applications as drug delivery vehicles for transporting even the efflux-pump vulnerable anticancer/ antimicrobial pharmaceuticals at the target site. The surface functionality prevents direct exposure of AgNPs; however, its loss results in toxicity. Therefore, the nature of interactions between AgNPs and surface ligand plays a significant role in AgNPs toxicity. Strong covalent interactions between the surface ligand and AgNPs result in reduced toxicity, whereas the weaker, physical interactions manifest deleterious effects. The development of AgNP-formulations, therefore, requires a consideration of toxicity, engineered by using a suitable surface functionalization; in addition to the identification of specific interactions between the nanoparticle and biomolecule.

2. SURFACE FABRICATION OF AgNPs

The chemically synthesized AgNPs, when internalized in the cells, trigger the conformational distortions in the surrounding cellular proteins following the loss of surface coatings and developing protein corona [1] on their surfaces that covers the fabricated surface and hinders the purpose of surface functionalities. These events cause the aggregation of AgNPs [2] and induce denaturation and structural distortions in proteins [3], hence exposing the cryptic peptide epitopes that instigate the autoimmune response [4]. Similarly, the silver ions furnished by the oxidation of AgNPs cause the production of intracellular ROS, thereby causing functional anomalies to the vital cell organelles and biomacromolecules [5]. In addition, the exposed AgNPs incapacitate several critical cell-signaling pathways and restrict the cells in S-phase and G1-phase that causes uncontrolled proliferation and apoptosis [6].

The presence of surface functionalization held by strong interactions with nanoparticle surface affords thermodynamic stability for evading the development of surface corona, which eventually ameliorates the nanoparticle stability *in vivo*, holding the original surface engineered physicochemical properties [7]. The forces such as steric stabilization, hydration forces, deletion stabilization, van der Waals forces, and electrostatic stabilization cater to the stabilization of AgNPs [8]. Similarly, the presence of surface functionalization extends the drug delivery applications of AgNPs while affording the sustained release of drug molecules at the target site [9]. Importantly, the tethering of nanoparticle surface with organic ligands affords supplementary functionalities that improve the compatibility with another phase and prevent the aggregation of nanoparticles in the colloidal phase [10]. Notably, the organic ligands accelerate the anchoring of AgNPs onto the support system, hence causing a high dispersion of the metallic nanoparticles in the target system [11]. Therefore, the capping agent determines the physicochemical and biological properties of AgNPs for offering various applications. Figs. (**1** and **2**) highlights the types of interactions between functionalized AgNPs and various biomolecules.

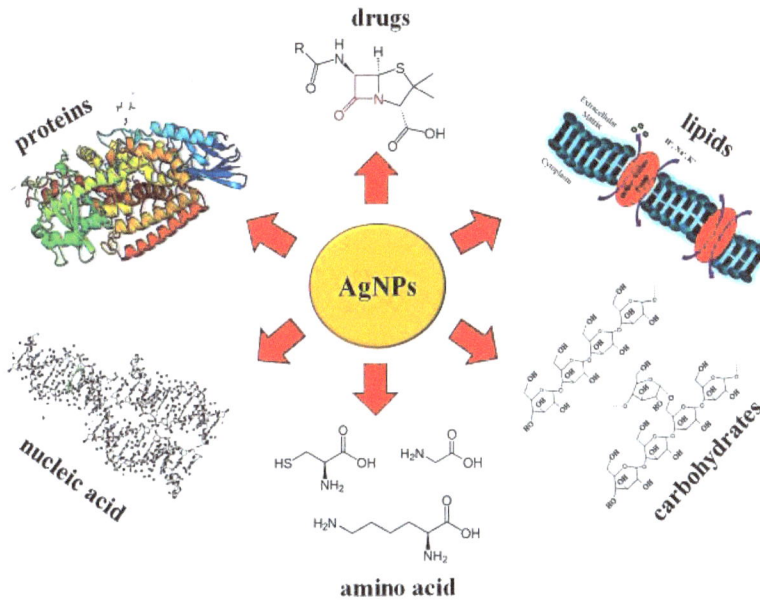

Fig. (1). Interactions of AgNPs with various biomolecules or bioactive molecules.

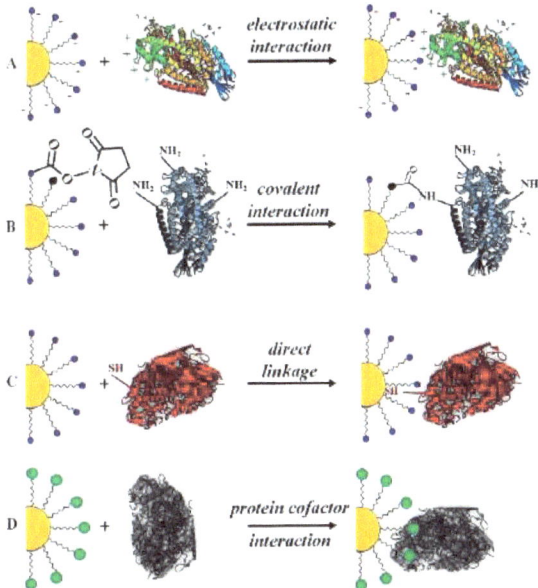

Fig. (2). Interactions of functionalized AgNPs with biomolecules.

2.1. Fabrication with Peptides

Conjugation of AgNPs with peptides proceeds through the approaches: electrostatic interactions, covalent linkage, direct linkage of the functional head group present on peptide to the nanoparticle surface and tethering of a protein cofactor on the nanoparticle surface [12]. Mainly, the RGD (arginylglycylaspartic acid) is responsible for cell adhesion to the extracellular matrix, TAT-like peptides and cell penetrating TAT (trans-activating transcriptional activator) conjugate with AgNPs for therapeutic applications [13, 14]. The cell-penetrating peptides provide the delivery of cargo drug molecules across the transmembrane pores into the subcellular compartments [15, 16]. On conjugation to AgNPs, the cell penetrating peptides offer applications in cancer therapy, drug delivery, neurology, intracellular delivery and many others [17]. Pal *et al.* 2016, investigated the involvement of cysteine residues in the interaction of a peptide with AgNPs and reported a weak bonding leading to a dynamic exchange of peptides from the nanoparticle surface without experiencing any conformational alterations [18]. Principally, the antimicrobial peptides (AMP) where the structure [19] and charge play a critical role in bioactivity display the weak interactions that can upset the AMP structure and ease their activity [20]. Importantly, the steric repulsions between AgNP-bioconjugated proteins restrain the nanoparticle aggregation and enhances the stability of the conjugate that assists in a multiple times recycling of AgNP-AMP conjugate. The investigations also proposed interactions between the peptide and the negatively charged phosphate head groups of lipid moieties as well as with water molecules. The protein nanoconjugates interact with the acyl chains of phospholipids of the cell membrane *via* hydrophobic interactions that act as a driving force for the activity of nanoconjugate. The conjugated peptide detaches from the AgNP surface, thereby forming a pore in the membrane [21]. The pore formation augments the bactericidal action of nanoconjugates by anchoring to the microbial DNA. Interestingly, the modified AMP shows a superior activity compared to the unmodified parent peptide because the binding of the former and its corresponding nano-conjugates with membranes is an energetically favorable process [22]. The peptide conjugated AgNPs also serve as favorable materials for targeting MDR strains of bacteria *Pseudomonas aeruginosa, Enterococcus faecium, Klebsiella pneumoniae, Staphylococcus aureus, Enterobacter species, and Acinetobacter baumannii,* by tagging a cysteine residue to at the N- or C-terminal of the parent antimicrobial peptides (AMPs): AY1C [FLPKLFAKITKKNMAHIRC], CAY1 [CFLPKLFAKITKKNMAHIR], and AY1 [FLPKLFAKITKKNMAHIR]. The resulting nanoparticle-peptide system displayed a minimum inhibitory concentration (MIC) in the range 5-15 µM with a superior antimicrobial activity as compared to AgNPs and AMPs taken together.

The AgNPs display enhanced stability in the presence of peptides and assist in the delivery of a large number of peptide molecules in the proximity of the bacterial cell membrane, leading to membrane rupture [23]. Mohanty *et al.* 2013, reported the antimicrobial activity of AgNPs decorated with cationic AMPs, such as NK-2, LLKKK-18, and LL37, against *Mycobacterium marinum* and *Mycobacterium smegmatis.* Reportedly, NK-2 and LLKKK-18 displayed an improved antimicrobial effect with biogenic AgNPs at the concentration 0.5-ppm against *M. smegmatis.* However, no synergistic effect appeared for NK-2 conjugated AgNPs against *M. marinum.* The biogenic AgNps also displayed substantial synergism with the antituberculosis drug rifampin against *M. smegmatis.* Furthermore, the AMPs-fabricated AgNPs displayed negligible cytotoxicity and DNA damage [24]. Avila *et al.* (2017) investigated oligodynamic properties of Ubiquicidin conjugated to AgNPs against *Escherichia coli* and *Pseudomonas aeruginosa.* A significant increase in the antimicrobial effect occurred at 75 μg/ mL due to an increase in the local concentration of peptide surrounding the nanoparticle nucleus distributed in a multimeric or polyvalent arrangement and the inherent biocidal effect of the AgNPs, including the release of silver ions from the nanoparticle core [25]. Farkhani *et al.* (2017) reported enhanced anticancer activity of cationic cell penetrating peptides (CPP)-anchored AgNPs [26]. The CPPs that act as intracellular transit vehicles [27] consists of positively charged amino acid residues, such as lysine and arginine [28]. Their conjugation with AgNPs lowers the negative potential of nanoparticles, thereby reducing the electrostatic barrier of the cell membrane [26]. Lambadi *et al.* (2015) conjugated AgNPs with polymyxin B for investigating the potency of the nanosystem against MDR strains of *P. aeruginosa* and *Vibrio fluvialis.* The test nanosystem achieved a three-fold higher oligodynamic effect as compared to the citrate-capped AgNPs. The reported nanoconjugates caused morphological damages to the microbial membrane and inhibited the biofilm formation [29]. Chaudhary *et al.* (2016) conjugated single walled carbon-nanotubes (SWCNT) to AgNPs and further linked them with AMP TP359. The test nanosystem displayed antimicrobial properties against gram-negative pathogens (*Salmonella enterica serovar Typhimurium* and *Escherichia coli*) and gram-positive (*Staphylococcus aureus* and *Streptococcus pyogenes*) bacteria. In addition, the reported nanosystem displayed lower toxicity towards the murine macrophages and Hep2 cells [30]. Bajaj *et al.* (2018) further conjugated AgNPs with small peptides *via* carbodiimide cross-linker chemistry, in the presence of mercaptopropanoic acid that serves as a cross-linker. The conjugates displayed significant antimicrobial activity against *Escherichia coli and Candida albicans* through cell wall disruption with a better inhibition profile as compared to the non-conjugated AgNPs [31]. Alghrair *et al.* (2019) recognized the antiviral bioactivity of AgNPs bioconjugate to peptide FluPep, regarded as an established inhibitor of influenza virus in model systems; against influenza A virus subtypes, including H1N1, H3N2 and H5N1. The FluPep peptide WLVFFVIFYFFRRRKK, appended with the sequence CVVVTAAA- at its N-terminal to get FLuPep ligand displayed an enhanced antiviral activity at low grafting densities on conjugation with AgNPs, which also assisted in the intracellular delivery of the peptide [32]. Reportedly, the FluPep functionalized AgNPs inhibited the viral plaque formation in canine MDCK cells with an IC50 2.1 nM [33]. Liu *et al.* 2012, reported AgNPs fabricated with TAT cell-penetrating peptides for targeting the MDR cancer. The test nanosystem presented marked antitumor activity in the MDR susceptible cells. Interestingly, the nanoconjugate exhibited 24-fold improvement in the antitumor effect as compared to doxorubicin [34]. Kittler *et al.* (2010) developed spherical shaped AgNPs with diameter 50 ± 20 nm stabilized with poly(N-vinylpyrrolidone) (PVP) and dispersed in different cell culture media: pure RPMI, RPMI containing up to 10% of bovine serum albumin (BSA), and RPMI containing up to 10% of fetal calf serum (FCS) to study the ultimate fate of nanoparticles in cells. Reportedly, the nanoparticles displayed a decrease in toxicity against human mesenchymal stem cells even at low BSA or FCS concentrations with FCS, exhibiting a more noticeable effect. The free silver ions released by AgNPs that contribute to physiological toxicity bind to BCA and FCS, thereby reducing direct cell toxicity [35]. Recently, Higa *et al.* 2019, developed spherical morphology, low polydispersity, face-centered cubic crystal structure, and an average size of 29.3 ± 3.0 nm and conjugated them to MOG (myelin oligodendrocyte glycoprotein) and MBP (Myelin basic protein) peptides. The bioconjugated nanosystem acted as a diagnostic probe for demyelinating diseases and autoantibodies. Variations in the hydrodynamic diameters revealed a grander agglomeration induced by MBP as compared to MOG peptides indicating the efficient binding of peptides to the AgNPs that upheld their bioactivity and assisted in the autoantibody recognition. These findings validate the suitability of peptide bioconjugated AgNPs in identifying the diagnostic biomarkers in the demyelination perspective [36].

2.2. Fabrication with Saccharides

Polysaccharides, regarded as outstanding ligands for the synthesis of bioconjugated AgNPs display a benign and steady biodegradation coupled with an admirable biocompatibility [37]. Notably, the host polysaccharides can accumulate a carrier with metallic ions and hydrophobic chemical drugs such as doxorubicin, ceftriaxone, ciprofloxacin, cefotaxime, and levofloxacin for targeted drug delivery [38]. The polysaccharide conjugated AgNPs exhibit outstanding applications in several arenas owing to their remarkable physical, chemical and biological properties that include targeted drug delivery, biosensing of essential metabolites, catalysis and nanopharmaceuticals with antimicrobial, antiviral and anticancer capabilities [39]. Sanyasi *et al.* 2016, developed CMT (carboxymethyl tamarind) polysaccharide capped AgNPs and tested their potency against both Gram positive (*Bacillus subtilis*) and Gram negative (*Escherichia coli* and *Salmonella typhimurium*) bacterial strains and MDR strains of *Staphlococcus haemolyticus, S. epidermidis, Escherichia coli C19, Klebsiella pneumoniae Kp52,* and *Enterobacter cloacae Ec18.* The CMT capped AGNPs displayed a dose-dependent antibacterial effect with MIC = 1.5 to 6 µg/ mL mainly by inhibiting the formation of biofilms and blocking the expression of FtsZ and FtsA, thereby arresting bacterial cell proliferation. Importantly, the CMT-capped AgNPs displayed the least cytotoxicity when tested for their cell viability on mouse macrophage RAW 264.7 cells, neuronal cells (such as F11), human keratinocytes (such as HaCaT cells) and human osteoblasts (such as SaOS cells) [40]. Further, Mishra *et al.* (2018) revealed that CMT-capped AgNPs significantly modulated the expression of transporter AcrB protein that results in overcoming MDR in *Enterobacter cloacae* isolates and their antibacterial potency was not inhibited by AcrAB-TolC efflux protein expression [41]. Dini *et al.* (2011) evaluated the *in vitro* cytotoxicity of glycans-capped AgNPs. Reportedly, the toxicity of capped AgNPs on tumor cells increased with incubation time and intracellular concentration due to the production of ROS that acts as signal molecules to promote cell cycle progression by affecting growth factor receptors, AP-1 and NFkB thereby inducing oxidative DNA damage [42]. Venkatesan *et al.* (2018) developed nanopharmaceutical based on AgNPs conjugated to chitosan-fucoidan polysaccharides by a strong polyelectrolyte complexation. The complex notably inhibited the growth of Gram-positive *staphylococcus aureus* and Gram-negative *Escherichia coli,* in addition to a significant anticancer activity in human cervical cancer cells (HeLa) [43]. The polysaccharides endowed with a porous structure and twisted conformations [44] of the subunits form an exciting material for capping AgNPs to function effectually as a drug delivery vehicle [45]. The vector effectively prompted an intracellular transportation of commercially available, mechanistically established but MDR vulnerable antibiotics [46, 47] by evading the active microbial efflux. Essentially, single helical V-amylose exhibits a brilliant tendency to form the helical inclusion complexes [48] with linear alcohols, fatty acids [49] and aromatic compounds [50]. Mainly, the host molecule accommodates between the helical-structure of amylose *via* H-bonding. In aqueous medium, amylose adopts helical conformation with its –OH groups pointing out the helix. Whereas, the glycosidic –O- and -CH$_2$- groups point towards the inner core of helix, hence creating a hydrophobic cavity. The hydrophobic and van der Waals interactions form the basis of complexation in the inclusion complexes of amylose. The acetylated amylose reportedly acts as an inclusion host for drugs such as rifampicin [51], ibuprofen [52], and indomethacin

[53], eventually improving the activity profile of these drugs. Heparin, a highly acidic, linear polysaccharide, exhibits numerous biological functions such as blood anticoagulation, anti-inflammation promotion of cell adhesion and cell migration [54]. For conjugating with AgNPs, 2,6-diaminopyridine (DAP) modifies the reducing end of heparin to yield DAPHP that binds tightly to the nanoparticles. The AgNP-DAPHP conjugates display notable anticoagulant and anti-inflammatory activities [55]. Additionally, Ag-DAPHP conjugates presented effective antimicrobial activity against *Staphylococcus aureus* and *Escherichia coli*, where DAPHP itself had no activity against *S. aureus* or *E. coli* [56] along with a sizeable anti-angiogenesis efficacy as validated by the inhibition of basic fibroblast growth factor-induced angiogenesis [57]. Similarly, hyaluronic acid that is a linear, high molecular weighed polydisperse polysaccharide, comprises of repeating disaccharide units of glucuronic acid and N-acetyl glucosamine residues. The polysaccharide deprived of any sulfo groups displays substantial biological compatibility and useful rheological properties [58]. The nanocomposites based on AgNPs with sizes ranging from 5 - 30 nm stabilized by hyaluronic acid also exhibited potent antimicrobial activity against *Staphylococcus aureus* and *Escherichia coli*. Chitin, a polysaccharide comprising β(1,4)-linked N-acetyl-D-glucosamine produces chitosan by removing the N-acetyl groups from chitin that displays better solubility, and the exposed primary amine functionalities on its surface effectively supports the immobilization of metallic nanoparticles. Carboxymethyl chitosan obtained by the carboxymethylation of hydroxyl and amino groups of chitosan acts as a matrix material for AgNPs [59]. The bioconjugates obtained by linking AgNPs with chitosan demonstrate an exceptional catalytic activity with multi-usability up to seven cycles [60]. Further, the extended application for the development of AgNP-chitosan scaffolds proved beneficial for wound dressings and for grafting onto various bio-implants by using cross- linking agents such as *Genipin* that improves the structural reinforcement and antibacterial properties of Ag-chitosan nanoparticle bioconjugates [61]. Hence, the AgNPs bioconjugated to polysaccharide serve as good candidates for use in medical applications.

2.3. Fabrication with Nucleic Acids

AgNPs bioconjugated with nucleic acids display profound applications in nano-biotechnology, cell imaging and drug delivery. Liang *et al.* validated a stress-free cellular transit of DNA functionalized nanomaterials than free single-stranded DNA for improved clinical applications. Reportedly, the DNA-templated *in situ* AgNPs synthesis offers effective drug vehicles [62]. Liu *et al.* (2015) designed DNA templated-AgNPs anchored on mesoporus silica nanospheres (MSNs) for the release of loaded drug, controlled by intracellular glutathione (GSH) [63].

Sarkar *et al.* (2015) constructed biofunctionalized AgNPs with superior DNA binding efficiency tailored for transfection [64]. On their immobilization to the Arg–Gly–Asp–Ser (RGDS) peptide, the transfection efficiency of AgNPs improved by 42 4% and 30 3% in HeLa and A549 cells, respectively. Tang *et al.* (2015) performed DNA-templated synthesis of AgNPs for the development of two-photon excitation (TPE) based sensors applicable for the detection of GSH in live cells. The conjugates demonstrated two-photon-sensitized fluorescence properties, in addition to delivering a superior cell permeability, and appreciable biocompatibility. The test conjugates selectively detected GSH even under the complex biological settings [65]. Sun *et al.* (2019) designed DNA-templated silver nanoparticles (AgNPs) with electrochemical atom transfer radical polymerization signal amplification as highly sensitive probes for the detection of DNA [66]. Lubitz *et al.* (2011) reported the biological properties of AgNPs conjugated to G-quadruplex DNA. AgNPs with size 20 nm tether to G-quadruplexes containing phosphorothioate anchor residues at both ends of the DNA. The conjugates consist of stacked G-tetrad plains for an effective π overlap compared to the base pairs of the canonical double-stranded DNA. The addition of AgNPs to the G-quadruplex ends afford better electric communication *via* metal/metal contact interactions, thereby supporting the DNA mediated charge migration [67]. Lee *et al.* (2007) reported novel approach for synthesizing cyclic disulfide-anchoring group tethered DNA based AgNP-oligonucleotide conjugates. These systems, upon functionalization with complementary sequences, aggregate reversibly to generate DNA-linked nanoparticle networks. These nanosystems qualify mainly for developing synthons in programmable materials synthesis approaches, molecular diagnostic labeling probes, and as functional components for plasmonic and nanoelectronic devices [68]. Komal *et al.* (2019) reported the interactions of AgNPs with Calf thymus DNA (CT-DNA) predominantly to be through van der Waals and H-bonding with the DNA groove leading to its destabilization. The bioconjugated AgNP mediated DNA destabilization may affect DNA replication, repair and transcription by lowering down the energy required to open up DNA helix for the development of anticancer nanopharmaceuticals [69]. Thompson *et al.* (2008) reported Oligonucleotide-AgNPs conjugates as probes for detecting a lower concentration of a specific DNA sequence due to their larger extinction coefficient [70]. Nasirian *et al.* (2017) developed a fluorescence resonance energy transfer (FRET)-based nanobioprobe for selective detection of aflatoxin B1 (AFB1). In the absence of AFB1, the aptamer and cDNA conjugate (aptamer-cDNA) undergo complexation, thereby bringing the polymer dots near AgNPs, resulting in the occurrence of FRET. Contrarily, the presence of AFB1 causes the aptamer release from cDNA-AgNP aggregates, resulting in fluorescence recovery [71]. Zheng *et al.* (2012) designed AgNP-DNA bionano-conjugates appended with a distinct number of

DNA ligands. The stabilization of the conjugates appeared to be valence controlled and gel electrophoresis verified their stability. The conjugation procedure strictly avoided the use of thiols and surfactants for the surface inactivity of AgNPs that leads to a sustained activity of nanoconjugates surface-dependent applications such as catalysis [72]. Pal *et al.* (2009) performed surface functionalization of AgNPs with chimeric phosphorothioate modified DNA (ps-po-DNA). The conjugates reportedly stabilize in buffer conditions pliable to DNA hybridization, with plausible applications in nanophotonics and biosensing [73]. Qi *et al.* (2017) developed origami paper analytical devices modified by (oPADs)raman-active, DNA-encoded anisotropic nanoparticles for a rapid, highly sensitive, and specific miRNA detection. The conjugates were prepared using 10-mer oligo-A, -T, -C, and -G for mediating the growth of Ag cubic seeds into Ag nanoparticles (AgNPs) with varying morphologies that were eventually encoded with DNA probes serving as operative surface-enhanced Raman scattering (SERS) probes. The economical, yet robust analytical device broadly applies to a range of miRNAs with a detection limit of 1 pM and assay time within 15 min, and hence warrants encouraging applications in point-of-care diagnostics [74]. Liu *et al.* (2012) reported the facile synthesis of DNA-AgNP conjugates for the quantitative detection of HIV DNA with a sandwich strategy based on their strong plasmon resonance scattering signals. These conjugates apply in the detection of low concentration of a specific DNA sequence based on the LSPR light scattering signals of AgNPs and for bioimaging [75]. Zhu *et al.* (2015) designed DNA-decorated AgNPs by a facile approach to self-assemble unmodified DNA on AgNPs by exploiting intrinsic silver-cytosine (Ag-C) coordination. The strong coordination readily promotes the DNA-AgNPs conjugation with promising stability under high ionic strength and high temperature. Interestingly, the reported nanoconjugates possess a highly efficient molecular recognition efficiency and fast hybridization kinetics compared to the thiolated DNA-modified AgNPs. The potential applications of these nanoconjugates involve plasmonics, DNA nanotechnology and the development of biosensors [76]. Abbaspour *et al.* (2014) reported sensitive and highly selective sandwich immunosensors for the detection of *Staphylococcus aureus*. In the reported bioassay, the capture probe was served by a biotinylated primary anti-*S.aureus* aptamer was immobilized on streptavidin coated magnetic beads (MB). The reported electrochemical immunosensor exhibited an extended dynamic range with a low detection limit [77]. Friedman *et al.* (2015) demonstrated a facile transcription of a fGmH (2′-F-dG, 2′-OMe-dA/dC/dU) RNA library with surprising hydrophobicity. The fGmH RNA aptamers substantially functionalize, stabilize, and further deliver aggregation-prone silver nanoparticles (AgNPs) to *S. aureus* with SpA-dependent antimicrobial effects. These conjugates possess a considerable potential to improve the *in vivo* applicability of nucleic acid-based affinity molecules to biomaterials, including deliberated smartness to biomaterials, and for antimicrobial AgNP delivery to *S. aureus* cells in a SpA-dependent manner [78]. Li *et al.* (2015) developed theranostic agent Ag-Sgc--FAM for anticancer therapy and the realization of fluorescence-enhanced cell imaging. The reported theranostic agent is readily internalized into cancer cells by receptor-mediated endocytosis. The reported aptamer-AgNPs conjugates displayed commendable potential as theranostic agents for inducing specific apoptosis of cells and achieving cells imaging in real time [79].

2.4. Fabrication with Amino Acids

The presence of diverse functional groups together with a good biocompatibility make amino acids as the molecules of interest in contemporary medicine. The bioconjugation of amino acids to AgNPs presents interesting clinical applications. Mondal *et al.* (2014) described the synthesis of AgNPs by an *in situ* reduction of silver nitrate with designed ethylenediaminetetraacetic acid-tryptophan where EDTA enhanced water solubility of the resulting bioconjugate. The conjugates dislayed a superior SERS activity that assisted in the high sensitivity detection of biomolecules. In addition, these bioconjugated AgNPs displayed an excellent catalytic activity towards the reduction of p-nitrophenol to p-aminophenol in water with a rate constant of 16.43×10^{-2} min^{-1} [80]. Dojcilovic *et al.* (2017) studied the interactions of the tryptophan functionalized AgNPs and live cells of *Candida albicans* synchrotron excitation deep-ultraviolet (DUV) fluorescence imaging at the DISCO beamline of Synchrotron SOLEIL. The DUV imaging explained that incubation of the fungus with functionalized AgNPs results in a significant increase in the fluorescence signal. The image-analysis revealed a less pronounced interaction of AgNPs with (pseudo)hyphae polymorphs of the diploid fungus in the case of yeast cells or budding spores. Due to a significant environmental sensitivity of the nano-bioconjugates, the plausible accumulation sites of nanoparticles could be readily determined. The intensity decay kinetic experiments described highly pronounced photobleaching effects in the case of the functionalized nanoparticle treated cells. Moreover, the time-integrated emission investigations proposed the cellular penetration of the nanoparticles; however, the majority of the nanoparticles attach to the cells' surfaces [81]. Chandra *et al.* (2018) described the biosynthesis of amino acid functionalized silver nanoparticles (AgNPs) using Neem gum by the green method. The amino acid – AgNP conjugates displayed a superior colloidal stability with surfactants and dyes. Reportedly, the tryptophan functionalized AgNPs showed the highest MB reduction rate in anionic micellar medium and the nanoparticles held responsible for the backward oxidation of reduced MB in an acidic nonionic micellar medium. Interestingly, the clock reaction validated the prospective application of the conjugated AgNPs as low-cost oxygen leakage sensors in vacuum-packed food packages, and for creating an oxygen deficient environment in acidic aqueous and micellar media, where catalyst poisoning by O_2 prevails [82]. Matos *et al.* (2017) reported the green synthesis of AgNPs by using 21 amino acids. Photoreduction method assisted in the synthesis of silver nanoparticles with tryptophan and tyrosine, methionine, cystine and histidine to yield Spherical AgNPs with sizes in the range 15 - 30 nm. The investigations further extended the applications of these nanosystems in biological systems due to the biocompatibility of the amino acids and pH control, such as the *in situ* synthesis of nanoparticles or the functionalization of metal nanoparticles with amino acids/ proteins [83]. Bonor *et al.* (2014) developed lysine conjugated AgNPs of 5 nm size by a rapid and convenient batch method. The lysine bioconjugated AgNPs with a size less than 100 nm displayed a notable clinical potency and an optimal choice for drug delivery because of its intrinsic anti-platelet, anti-bacterial and anti-inflammatory capabilities of the amino acid. The bioactivity of nano-bioconjugates revealed limited toxicity in cells, using HEK 293 cell line as a model system [84]. Shankar *et al.* (2015) synthesized amino acids tyrosine and tryptophan capped AgNPs incorporated into the agar to prepare antimicrobial composite films. No change in the physicochemical properties such as chemical structure, thermal stability, moisture content, and water vapor permeability appeared upon the incorporation of amino acid capped AgNPs into the agar. In addition, the AgNPs-agar nanocomposite films displayed commendable antimicrobial activities, primarily against *Listeria monocytogenes* and *Escherichia coli*. The described nanocomposites display perspective applications to the active food packaging by controlling the food-borne pathogens [85]. Khan *et al.* (2012) reported the synthesis and surface modification of colloidal silver nanoparticles with the cysteine by electrochemically active biofilm (EAB) used as a reducing agent. Conjugation with amino acid provides amine and carboxylic acid functional groups on the surface of the AgNPs that lead to hydrogen bonding and cross-linking of the nanoparticles in aqueous solution. The biological investigations described that the Cys-AgNPs bioconjugates display superior antibacterial activity against *Escherichia coli* and *Pseudomonas aeruginosa* [86]. Kasthuri *et al.* (2009) reported a novel approach for synthesizing aqueous stable AgNPs by using naturally occurring amino acid conjugated sodium salt of taurocholate (NaTC) as a reducing agent and sodium salt of glycocholate (NaGC) as a capping agent. UV-vis spectroscopy kinetically monitored the AgNP formation. By altering the nature of bile salts, shape and size of AgNPs changed. Techniques such as FT-IR spectroscopy, cyclic voltammetry (CV) and thermogravimetry analysis (TGA) validated the interaction between nanoparticles with bile salts [87]. Daima *et al.* (2013) described the designing of silver based nanoconjugates by surface modification of AgNPs with the surface corona of biologically active polyoxometalates (POMs) such as phosphotungstic acid (PTA) and phosphomolybdic acid (PMA) by utilising zwitterionic tyrosine amino acid as a pH-switchable reducing and capping agent of AgNPs. The antibacterial investigations on gram-negative bacterium *Escherichia coli* and gram-positive bacterium *Staphylococcus albus* established that tyrosine conjugated AgNPs on the surface corona of POMs enhances the physical damage to the bacterial cells due to a synergistic biocidal effect of AgNPs and POMs. It also validated the capability of tyrosine-reduced AgNPs to act as an excellent carrier and stabilizer for the POMs [88]. These properties support that bioconjugation and surface functionalization of AgNPs provided suitable material for applications in biomedicine.

CONCLUSION

AgNPs possess several physical, optical, plasmonic, electric and thermal properties, which become masked during their applications in different fields. The non-functionalized AgNPs pose high toxicity to the biological systems and present heightened immunogenicity. Surface engineering of AgNPs with suitable moieties or functional groups prevents their direct contact with physiological settings, hence mitigating the disadvantages associated with a direct surface contact. The surface functionalization of AgNPs allows their further conjugation with molecules of interest that couple the therapeutic potency of the conjugate and the physicochemical properties of the nanoparticles for extended applications. Therefore, the surface conjugation of AgNPs with selected molecules provides the basis of biomolecular applications with minimal toxicity and immunogenicity.

REFERENCES

[1]   Bannunah, A.M.; Vllasaliu, D.; Lord, J.; Stolnik, S. Mechanisms of nanoparticle internalization and transport across an intestinal epithelial cell model: effect of size and surface charge. *Mol. Pharm.,* **2014**, *11*(12), 4363-4373.
       [http://dx.doi.org/10.1021/mp500439c] [PMID: 25327847]

[2]   Alla, P.K.; Lauf, P.K.; Pavel, I.; Paluri, A.; Marcopoulos, M.; Yaklic, J.; Adragna, N.C. Internalized silver nanoparticles alter ion transport and hemoglobin spectrum in human red blood cells. *FASEB J.,* **2016**, *30*, Ib620.

[3]   Zhang, L.; Cheng, H.; Zheng, C.; Dong, F.; Man, S.; Dai, Y.; Yu, P. Structural and release properties of amylose inclusion complexes with ibuprofen. *J. Drug Deliv. Sci. Technol.,* **2016**, *31*, 101-107.
       [http://dx.doi.org/10.1016/j.jddst.2015.12.006]

[4]   Haase, H.; Fahmi, A.; Mahltig, B. Impact of silver nanoparticles and silver ions on innate immune cells. *J. Biomed. Nanotechnol.,* **2014**, *10*(6), 1146-1156.
       [http://dx.doi.org/10.1166/jbn.2014.1784] [PMID: 24749409]

[5]   Prasher, P.; Sharma, M.; Mudila, H.; Gupta, G.; Sharma, A.K.; Kumar, D.; Bakshi, H.A.; Negi, P.; Kapoor, D.N.; Chellappan, D.K.; Tambuwala, M.M.; Dua, K. Emerging trends in clinical implications of bio-conjugated silver nanoparticles in drug delivery. *Colloid Interface Sci. Commun.,* **2020**, *35*: 100244.
       [http://dx.doi.org/10.1016/j.colcom.2020.100244]

[6]   Garcia, E.B.; Alms, E.; Hinman, A.W.; Kelly, C. Single cell analysis reveals that chronic silver nanoparticle exposure induces cell division defects in human epithelial cells. *Int. J. Environ. Res. Pub. Health.,* **2019**, *16*, Article 2061.

[7]   Devi, L.B.; Das, S.K.; Mandal, A.B. Impact of Surface Functionalization of AgNPs on Binding and Conformational Change of Hemoglobin (Hb) and Hemolytic Behavior. *J. Phys. Chem. C,* **2014**, *118*(51), 29739-29749.
[http://dx.doi.org/10.1021/jp5075048]

[8]   Abraham, A.N.; Sharma, T.K.; Bansal, V.; Shukla, R. Phytochemicals as dynamic surface ligands to control nanoparticle-protein interactions. *ACS Omega,* **2018**, *3*(2), 2220-2229.
[http://dx.doi.org/10.1021/acsomega.7b01878] [PMID: 30023827]

[9]   Burduşel, A-C.; Gherasim, O.; Grumezescu, A.M.; Mogoantă, L.; Ficai, A.; Andronescu, E. Biomedical Applications of Silver Nanoparticles: An Up-to-Date Overview. *Nanomaterials (Basel),* **2018**, *8*(9), 681.
[http://dx.doi.org/10.3390/nano8090681] [PMID: 30200373]

[10]   Pokhrel, L.R.; Dubey, B.; Scheuerman, P.R. Impacts of select organic ligands on the colloidal stability, dissolution dynamics, and toxicity of silver nanoparticles. *Environ. Sci. Technol.,* **2013**, *47*(22), 12877-12885.
[http://dx.doi.org/10.1021/es403462j] [PMID: 24144348]

[11]   Afshinnia, K.; Marrone, B.; Baalousha, M. Potential impact of natural organic ligands on the colloidal stability of silver nanoparticles. *Sci. Total Environ.,* **2018**, *625*, 1518-1526.
[http://dx.doi.org/10.1016/j.scitotenv.2017.12.299] [PMID: 29996448]

[12]   Aubin-Tam, M-E.; Hamad-Schifferli, K. Structure and function of nanoparticle-protein conjugates. *Biomed. Mater.,* **2008**, *3*(3): 034001.
[http://dx.doi.org/10.1088/1748-6041/3/3/034001] [PMID: 18689927]

[13]   Liu, E.; Zhang, M.; Cui, H.; Gong, J.; Huang, Y.; Wang, J.; Cui, Y.; Dong, W.; Sun, L.; He, H.; Yang, V.C. Tat-functionalized Ag-Fe$_3$O$_4$ nano-composites as tissue-penetrating vehicles for tumor magnetic targeting and drug delivery. *Acta Pharm. Sin. B,* **2018**, *8*(6), 956-968.
[http://dx.doi.org/10.1016/j.apsb.2018.07.012] [PMID: 30505664]

[14]   Pietro, P.D.; Zaccaro, L.; Comegna, D.; Getto, A.D.; Saviano, M.; Snyders, R.; Cossement, D.; Satriano, C.; Rizzarelli, E. Silver nanoparticles functionalized with a fluorescent cyclic RGD peptide: a versatile integrin targeting platform for cells and bacteria. *RSC Advances,* **2016**, *6*(113), 112381-112392.
[http://dx.doi.org/10.1039/C6RA21568H]

[15]   Muñoz-Morris, M.A.; Heitz, F.; Divita, G.; Morris, M.C. The peptide carrier Pep-1 forms biologically efficient nanoparticle complexes. *Biochem. Biophys. Res. Commun.,* **2007**, *355*(4), 877-882.
[http://dx.doi.org/10.1016/j.bbrc.2007.02.046] [PMID: 17331466]

[16]   Asgary, V.; Shoari, A.; Baghbani-Arani, F.; Sadat Shandiz, S.A.; Khosravy, M.S.; Janani, A.; Bigdeli, R.; Bashar, R.; Cohan, R.A. Green synthesis and evaluation of silver nanoparticles as adjuvant in rabies veterinary vaccine. *Int. J. Nanomedicine,* **2016**, *11*, 3597-3605.
[http://dx.doi.org/10.2147/IJN.S109098] [PMID: 27536101]

[17]   Banerjee, K.; Ravishankar Rai, V.; Umashankar, M. Effect of peptide-conjugated nanoparticles on cell lines. *Prog. Biomater.,* **2019**, *8*(1), 11-21.
[http://dx.doi.org/10.1007/s40204-019-0106-9] [PMID: 30661226]

[18]   Pal, I.; Brahmkhatri, V.P.; Bera, S.; Bhattacharyya, D.; Quirishi, Y.; Bhunia, A.; Atreya, H.S. Enhanced stability and activity of an antimicrobial peptide in conjugation with silver nanoparticle. *J. Colloid Interface Sci.,* **2016**, *483*, 385-393.
[http://dx.doi.org/10.1016/j.jcis.2016.08.043] [PMID: 27585423]

[19]   Zhao, J.; Zhao, C.; Liang, G.; Zhang, M.; Zheng, J. Engineering antimicrobial peptides with improved

antimicrobial and hemolytic activities. *J. Chem. Inf. Model.,* **2013**, *53*(12), 3280-3296.
[http://dx.doi.org/10.1021/ci400477e] [PMID: 24279498]

[20]   Ghosh, A.; Kar, R.K.; Jana, J.; Saha, A.; Jana, B.; Krishnamoorthy, J.; Kumar, D.; Ghosh, S.; Chatterjee, S.; Bhunia, A. Indolicidin targets duplex DNA: structural and mechanistic insight through a combination of spectroscopy and microscopy. *ChemMedChem,* **2014**, *9*(9), 2052-2058.
[http://dx.doi.org/10.1002/cmdc.201402215] [PMID: 25044630]

[21]   Brahmkhatri, V.P.; Chandra, K.; Dubey, A.; Atreya, H.S. An ultrastable conjugate of silver nanoparticles and protein formed through weak interactions. *Nanoscale,* **2015**, *7*(30), 12921-12931.
[http://dx.doi.org/10.1039/C5NR03047A] [PMID: 26166696]

[22]   Gupta, Y.K.; Chan, S-H.; Xu, S-Y.; Aggarwal, A.K. Structural basis of asymmetric DNA methylation and ATP-triggered long-range diffusion by EcoP15I. *Nat. Commun.,* **2015**, *6*(1), 7363.
[http://dx.doi.org/10.1038/ncomms8363] [PMID: 26067164]

[23]   Pal, I.; Bhattacharyya, D.; Kar, R.K.; Zarena, D.; Bhunia, A.; Atreya, H.S. A Peptide-Nanoparticle System with Improved Efficacy against Multidrug Resistant Bacteria. *Sci. Rep.,* **2019**, *9*(1), 4485.
[http://dx.doi.org/10.1038/s41598-019-41005-7] [PMID: 30872680]

[24]   Mohanty, S.; Jena, P.; Mehta, R.; Pati, R.; Banerjee, B.; Patil, S.; Sonawane, A. Cationic antimicrobial peptides and biogenic silver nanoparticles kill mycobacteria without eliciting DNA damage and cytotoxicity in mouse macrophages. *Antimicrob. Agents Chemother.,* **2013**, *57*(8), 3688-3698.
[http://dx.doi.org/10.1128/AAC.02475-12] [PMID: 23689720]

[25]   Avila, E.M.; Ferro-Flores, G.; Ocampo-García, B.E.; López-Téllez, G.; López-Ortega, J.; Rogel-Ayala, D-G.; Sánchez-Padilla, D. Antibacterial Efficacy of Gold and Silver Nanoparticles Functionalized with the Ubiquicidin Antimicrobial Peptide. *J. Nanomater.,* **2017**, *2017*: 5831959.

[26]   Mussa Farkhani, S.; Asoudeh Fard, A.; Zakeri-Milani, P.; Shahbazi Mojarrad, J.; Valizadeh, H. Enhancing antitumor activity of silver nanoparticles by modification with cell-penetrating peptides. *Artif. Cells Nanomed. Biotechnol.,* **2017**, *45*(5), 1029-1035.
[http://dx.doi.org/10.1080/21691401.2016.1200059] [PMID: 27357085]

[27]   Amand, H.L.; Rydberg, H.A.; Fornander, L.H.; Lincoln, P.; Nordén, B.; Esbjörner, E.K. Cell surface binding and uptake of arginine- and lysine-rich penetratin peptides in absence and presence of proteoglycans. *Biochim. Biophys. Acta,* **2012**, *1818*(11), 2669-2678.
[http://dx.doi.org/10.1016/j.bbamem.2012.06.006] [PMID: 22705501]

[28]   Heitz, F.; Morris, M.C.; Divita, G. Twenty years of cell-penetrating peptides: from molecular mechanisms to therapeutics. *Br. J. Pharmacol.,* **2009**, *157*(2), 195-206.
[http://dx.doi.org/10.1111/j.1476-5381.2009.00057.x] [PMID: 19309362]

[29]   Lambadi, P.R.; Sharma, T.K.; Kumar, P.; Vasnani, P.; Thalluri, S.M.; Bisht, N.; Pathania, R.; Navani, N.K. Facile biofunctionalization of silver nanoparticles for enhanced antibacterial properties, endotoxin removal, and biofilm control. *Int. J. Nanomedicine,* **2015**, *10*, 2155-2171.
[PMID: 25834431]

[30]   Chaudhary, A.A.; Ashmore, D.; Nath, S.; Kate, K.; Dennis, V.; Singh, S.R.; Owen, D.R.; Palazzo, C.; Arnold, R.D.; Miller, M.E.; Pillai, S.R. **2016**.

[31]   Bajaj, M.; Pandey, S.K.; Wangoo, N.; Sharma, R.K. Peptide Functionalized Metallic Nanoconstructs: Synthesis, Structural Characterization, and Antimicrobial Evaluation. *ACS Biomater. Sci. Eng.,* **2018**, *4*(2), 739-747.
[http://dx.doi.org/10.1021/acsbiomaterials.7b00729] [PMID: 33418761]

[32]   Alghrair, Z.K.; Fernig, D.G.; Ebrahimi, B. Enhanced inhibition of influenza virus infection by peptide-noble-metal nanoparticle conjugates. *Beilstein J. Nanotechnol.,* **2019**, *10*, 1038-1047.
[http://dx.doi.org/10.3762/bjnano.10.104] [PMID: 31165030]

[33]   Nicol, M.Q.; Ligertwood, Y.; Bacon, M.N.; Dutia, B.M.; Nash, A.A. A novel family of peptides with potent activity against influenza A viruses. *J. Gen. Virol.,* **2012**, *93*(Pt 5), 980-986.

[http://dx.doi.org/10.1099/vir.0.038679-0] [PMID: 22258859]

[34]   Liu, J.; Zhao, Y.; Guo, Q.; Wang, Z.; Wang, H.; Yang, Y.; Huang, Y. TAT-modified nanosilver for combating multidrug-resistant cancer. *Biomaterials,* **2012**, *33*(26), 6155-6161.
[http://dx.doi.org/10.1016/j.biomaterials.2012.05.035] [PMID: 22682937]

[35]   Kittler, S.; Greulich, C.; Gebauer, J.S.; Diendorf, J.; Treuel, L.; Ruiz, L.; Gonzales-Calbet, J.M.; Vallet-Regi, M.; Zellner, R.; Koller, M.; Epple, M. The influence of proteins on the dispersability and cell-biological activity of silver nanoparticles. *J. Mater. Chem.,* **2010**, *20*(3), 512-518.
[http://dx.doi.org/10.1039/B914875B]

[36]   Higa, A.M.; Mambrini, G.P.; Ierich, J.C.M.; Garcia, P.S.; Scramin, J.A.; Peroni, L.A.; Okuda-Shinagawa, N.M.; Teresa Machini, M.; Trivinho-Strixino, F.; Leite, F.L. Peptide-Conjugated Silver Nanoparticle for Autoantibody Recognition. *J. Nanosci. Nanotechnol.,* **2019**, *19*(12), 7564-7573.
[http://dx.doi.org/10.1166/jnn.2019.16734] [PMID: 31196262]

[37]   Wang, C.; Gao, X.; Chen, Z.; Chen, Y.; Chen, H. Preparation, Characterization and Application of Polysaccharide-Based Metallic Nanoparticles: A Review. *Polymers (Basel),* **2017**, *9*(12), 689.
[http://dx.doi.org/10.3390/polym9120689] [PMID: 30965987]

[38]   Ivanova, N.; Gugleva, V.; Dobreva, M.; Pehlivanov, I.; Stefanov, S.; Andonova, V. Silver Nanoparticles as Multi-Functional Drug Delivery Systems, Nanomedicines, Muhammad Akhyar Farrukh, IntechOpen. **2018**.
[http://dx.doi.org/10.5772/intechopen.80238]

[39]   Buttacavoli, M.; Albanese, N.N.; Di Cara, G.; Alduina, R.; Faleri, C.; Gallo, M.; Pizzolanti, G.; Gallo, G.; Feo, S.; Baldi, F.; Cancemi, P. Anticancer activity of biogenerated silver nanoparticles: an integrated proteomic investigation. *Oncotarget,* **2017**, *9*(11), 9685-9705.
[http://dx.doi.org/10.18632/oncotarget.23859] [PMID: 29515763]

[40]   Sanyasi, S.; Majhi, R.K.; Kumar, S.; Mishra, M.; Ghosh, A.; Suar, M.; Satyam, P.V.; Mohapatra, H.; Goswami, C.; Goswami, L. Polysaccharide-capped silver Nanoparticles inhibit biofilm formation and eliminate multi-drug-resistant bacteria by disrupting bacterial cytoskeleton with reduced cytotoxicity towards mammalian cells. *Sci. Rep.,* **2016**, *6*(1), 24929.
[http://dx.doi.org/10.1038/srep24929] [PMID: 27125749]

[41]   Mishra, M.; Kumar, S.; Majhi, R.K.; Goswami, L.; Goswami, C.; Mohapatra, H. Antibacterial Efficacy of Polysaccharide Capped Silver Nanoparticles Is Not Compromised by AcrAB-TolC Efflux Pump. *Front. Microbiol.,* **2018**, *9*, 823.
[http://dx.doi.org/10.3389/fmicb.2018.00823] [PMID: 29780364]

[42]   Dini, L.; Panzarini, E.; Serra, A.; Buccolieri, A.; Manno, D. Synthesis and *In Vitro* Cytotoxicity of Glycans-Capped Silver Nanoparticles. *Nanomater. Nanotechnol.,* **2011**, *1*, 58-64.
[http://dx.doi.org/10.5772/50952]

[43]   Venkatesan, J.; Singh, S.K.; Anil, S.; Kim, S-K.; Shim, M.S. Preparation, Characterization and Biological Applications of Biosynthesized Silver Nanoparticles with Chitosan-Fucoidan Coating. *Molecules,* **2018**, *23*(6), 1429.
[http://dx.doi.org/10.3390/molecules23061429] [PMID: 29895803]

[44]   Sujka, M.; Pankiewicz, U.; Kowalski, R.; Nowosad, K.; Noszczyk-Nowak, A. Porous starch and its application in drug delivery systems. *Polim. Med.,* **2018**, *48*(1), 25-29.
[http://dx.doi.org/10.17219/pim/99799] [PMID: 30657655]

[45]   Liu, Z.; Jiao, Y.; Wang, Y.; Zhou, C.; Zhang, Z. Polysaccharides-based nanoparticles as drug delivery systems. *Adv. Drug Deliv. Rev.,* **2008**, *60*(15), 1650-1662.
[http://dx.doi.org/10.1016/j.addr.2008.09.001] [PMID: 18848591]

[46]   Ismail, N.S.; Gopinath, S.C.B. Enhanced antibacterial effect by antibiotic loaded starch nanoparticle. *J. Assoc. Arab Univ. Basic Appl. Sci.,* **2017**, *24*(1), 136-140.
[http://dx.doi.org/10.1016/j.jaubas.2016.10.005]

[47]   Fontes, G.C.; Calado, V.M.A.; Rossi, A.M.; da Rocha-Leão, M.H. Characterization of antibiotic-loaded alginate-OSA starch microbeads produced by ionotropic pregelation. *BioMed Res. Int.,* **2013**, *2013*: 472626.
[http://dx.doi.org/10.1155/2013/472626] [PMID: 23862146]

[48]   Rondeau-Mouro, C.; Le Bail, P.; Buléon, A. Structural investigation of amylose complexes with small ligands: inter- or intra-helical associations? *Int. J. Biol. Macromol.,* **2004**, *34*(5), 309-315.
[http://dx.doi.org/10.1016/j.ijbiomac.2004.09.002] [PMID: 15556233]

[49]   Lalush, I.; Bar, H.; Zakaria, I.; Eichler, S.; Shimoni, E. Utilization of amylose-lipid complexes as molecular nanocapsules for conjugated linoleic Acid. *Biomacromolecules,* **2005**, *6*(1), 121-130.
[http://dx.doi.org/10.1021/bm049644f] [PMID: 15638512]

[50]   Biais, B.; Le-Bail, P.; Robert, P.; Pontoire, B.; Buleon, A. Structural and stoichiometric studies of complexes between aroma compounds and amylose. Polymorphic transitions and quantification in amorphous and crystalline areas. *Carbohydr. Polym.,* **2006**, *66*(3), 306-315.
[http://dx.doi.org/10.1016/j.carbpol.2006.03.019]

[51]   Ribeiro, A.C.; Rocha, Â.; Soares, R.M.D.; Fonseca, L.P.; da Silveira, N.P. Synthesis and characterization of acetylated amylose and development of inclusion complexes with rifampicin. *Carbohydr. Polym.,* **2017**, *157*, 267-274.
[http://dx.doi.org/10.1016/j.carbpol.2016.09.064] [PMID: 27987927]

[52]   Zhang, X-F.; Shen, W.; Gurunathan, S. Silver Nanoparticle-Mediated Cellular Responses in Various Cell Lines: An *in Vitro* Model. *Int. J. Mol. Sci.,* **2016**, *17*(10), 1603.
[http://dx.doi.org/10.3390/ijms17101603] [PMID: 27669221]

[53]   Cai, X. Targeted and Controlled Release of Indomethacin from a Prodrug of Amylose. In: Peng Y., Weng X. (Eds) 7th Asian-Pacific Conference on Medical and Biological Engineering. IFMBE Proceedings, vol 19. Springer, Berlin, Heidelberg, **2008**.
[http://dx.doi.org/10.1007/978-3-540-79039-6_8]

[54]   Laremore, T.N.; Zhang, F.; Dordick, J.S.; Liu, J.; Linhardt, R.J. Recent progress and applications in glycosaminoglycan and heparin research. *Curr. Opin. Chem. Biol.,* **2009**, *13*(5-6), 633-640.
[http://dx.doi.org/10.1016/j.cbpa.2009.08.017] [PMID: 19781979]

[55]   Kemp, M.M.; Kumar, A.; Mousa, S.; Park, T.J.; Ajayan, P.; Kubotera, N.; Mousa, S.A.; Linhardt, R.J. Synthesis of gold and silver nanoparticles stabilized with glycosaminoglycans having distinctive biological activities. *Biomacromolecules,* **2009**, *10*(3), 589-595.
[http://dx.doi.org/10.1021/bm801266t] [PMID: 19226107]

[56]   Kemp, M.M.; Kumar, A.; Clement, D.; Ajayan, P.; Mousa, S.; Linhardt, R.J. Hyaluronan- and heparin-reduced silver nanoparticles with antimicrobial properties. *Nanomedicine (Lond.),* **2009**, *4*(4), 421-429.
[http://dx.doi.org/10.2217/nnm.09.24] [PMID: 19505245]

[57]   Kemp, M.M.; Kumar, A.; Mousa, S. Gold and silver nanoparticles conjugated with heparin derivative possess antiangiogenesis properties. *Nanotechnology,* **2009**, *20*, Article ID. 455104.

[58]   Doughty, M.J.; Glavin, S. Efficacy of different dry eye treatments with artificial tears or ocular lubricants: a systematic review. *Ophthalmic Physiol. Opt.,* **2009**, *29*(6), 573-583.
[http://dx.doi.org/10.1111/j.1475-1313.2009.00683.x] [PMID: 19686307]

[59]   Laudenslager, M.J.; Schiffman, J.D.; Schauer, C.L. Carboxymethyl chitosan as a matrix material for platinum, gold, and silver nanoparticles. *Biomacromolecules,* **2008**, *9*(10), 2682-2685.
[http://dx.doi.org/10.1021/bm800835e] [PMID: 18816099]

[60]   Thomas, V.; Yallapu, M.M.; Sreedhar, B.; Bajpai, S.K. Fabrication, characterization of chitosan/nanosilver film and its potential antibacterial application. *J. Biomater. Sci. Polym. Ed.,* **2009**, *20*(14), 2129-2144.
[http://dx.doi.org/10.1163/156856209X410102] [PMID: 19874682]

[61]   Sathishkumar, M.; Sneha, K.; Won, S.W.; Cho, C.W.; Kim, S.; Yun, Y.S. Cinnamon zeylanicum bark extract and powder mediated green synthesis of nano-crystalline silver particles and its bactericidal activity. *Colloids Surf. B Biointerfaces,* **2009**, *73*(2), 332-338.
[http://dx.doi.org/10.1016/j.colsurfb.2009.06.005] [PMID: 19576733]

[62]   Liang, H.; Zhang, X.B.; Lv, Y.; Gong, L.; Wang, R.; Zhu, X.; Yang, R.; Tan, W. Functional DNA-containing nanomaterials: cellular applications in biosensing, imaging, and targeted therapy. *Acc. Chem. Res.,* **2014**, *47*(6), 1891-1901.
[http://dx.doi.org/10.1021/ar500078f] [PMID: 24780000]

[63]   Liu, C.; Qing, Z.; Zheng, J.; Deng, L.; Ma, C.; Li, J.; Li, Y.; Yang, S.; Yang, J.; Wang, J.; Tan, W.; Yang, R. DNA-templated *in situ* growth of silver nanoparticles on mesoporous silica nanospheres for smart intracellular GSH-controlled release. *Chem. Commun. (Camb.),* **2015**, *51*(30), 6544-6547.
[http://dx.doi.org/10.1039/C5CC00557D] [PMID: 25765340]

[64]   Sarkar, K.; Banerjee, S.L.; Kundu, P.P.; Madras, G.; Chatterjee, K. Biofunctionalized surface-modified silver nanoparticles for gene delivery. *J. Mater. Chem. B Mater. Biol. Med.,* **2015**, *3*(26), 5266-5276.
[http://dx.doi.org/10.1039/C5TB00614G] [PMID: 32262602]

[65]   Tang, Q.; Wang, N.; Zhou, F.; Deng, T.; Zhang, S.; Li, J.; Yang, R.; Zhong, W.; Tan, W. A novel AgNP/DNA/TPdye conjugate-based two-photon nanoprobe for GSH imaging in cell apoptosis of cancer tissue. *Chem. Commun. (Camb.),* **2015**, *51*(94), 16810-16812.
[http://dx.doi.org/10.1039/C5CC06471F] [PMID: 26435127]

[66]   Sun, H.; Kong, J.; Wang, Q.; Liu, Q.; Zhang, X. Dual Signal Amplification by eATRP and DNA-Templated Silver Nanoparticles for Ultrasensitive Electrochemical Detection of Nucleic Acids. *ACS Appl. Mater. Interfaces,* **2019**, *11*(31), 27568-27573.
[http://dx.doi.org/10.1021/acsami.9b08037] [PMID: 31313584]

[67]   Lubitz, I.; Kotlyar, A. Self-assembled G4-DNA-silver nanoparticle structures. *Bioconjug. Chem.,* **2011**, *22*(3), 482-487.
[http://dx.doi.org/10.1021/bc1004872] [PMID: 21319752]

[68]   Lee, J-S.; Lytton-Jean, A.K.R.; Hurst, S.J.; Mirkin, C.A. Silver nanoparticle-oligonucleotide conjugates based on DNA with triple cyclic disulfide moieties. *Nano Lett.,* **2007**, *7*(7), 2112-2115.
[http://dx.doi.org/10.1021/nl071108g] [PMID: 17571909]

[69]   Komal, ; Sonia, ; Kukreti, S.; Kaushik, M. Exploring the potential of environment friendly silver nanoparticles for DNA interaction: Physicochemical approach. *J. Photochem. Photobiol. B,* **2019**, *194*, 158-165.
[http://dx.doi.org/10.1016/j.jphotobiol.2019.03.022] [PMID: 30954875]

[70]   Thompson, D.G.; Enright, A.; Faulds, K.; Smith, W.E.; Graham, D. Ultrasensitive DNA detection using oligonucleotide-silver nanoparticle conjugates. *Anal. Chem.,* **2008**, *80*(8), 2805-2810.
[http://dx.doi.org/10.1021/ac702403w] [PMID: 18307361]

[71]   Nasirian, V.; Chabok, A.; Barati, A.; Rafienia, M.; Arabi, M.S.; Shamsipur, M. Ultrasensitive aflatoxin B1 assay based on FRET from aptamer labelled fluorescent polymer dots to silver nanoparticles labeled with complementary DNA. *Mikrochim. Acta,* **2017**, *184*(12), 4655-4662.
[http://dx.doi.org/10.1007/s00604-017-2508-5]

[72]   Zheng, Y.; Li, Y.; Deng, Z. Silver nanoparticle-DNA bionanoconjugates bearing a discrete number of DNA ligands. *Chem. Commun. (Camb.),* **2012**, *48*(49), 6160-6162.
[http://dx.doi.org/10.1039/c2cc32338a] [PMID: 22588332]

[73]   Pal, S.; Sharma, J.; Yan, H.; Liu, Y. Stable silver nanoparticle-DNA conjugates for directed self-assembly of core-satellite silver-gold nanoclusters. *Chem. Commun. (Camb.),* **2009**, (40), 6059-6061.
[http://dx.doi.org/10.1039/b911069k] [PMID: 19809643]

[74]   Qi, L.; Xiao, M.; Wang, X.; Wang, C.; Wang, L.; Song, S.; Qu, X.; Li, L.; Shi, J.; Pei, H. DNA-

Encoded Raman-Active Anisotropic Nanoparticles for microRNA Detection. *Anal. Chem.*, **2017**, *89*(18), 9850-9856.
[http://dx.doi.org/10.1021/acs.analchem.7b01861] [PMID: 28849911]

[75]   Liu, Y.; Huang, C.Z. One-step conjugation chemistry of DNA with highly scattered silver nanoparticles for sandwich detection of DNA. *Analyst (Lond.)*, **2012**, *137*(15), 3434-3436.
[http://dx.doi.org/10.1039/c2an35167f] [PMID: 22669124]

[76]   Zhu, D.; Chao, J.; Pei, H.; Zuo, X.; Huang, Q.; Wang, L.; Huang, W.; Fan, C. Coordination-mediated programmable assembly of unmodified oligonucleotides on plasmonic silver nanoparticles. *ACS Appl. Mater. Interfaces*, **2015**, *7*(20), 11047-11052.
[http://dx.doi.org/10.1021/acsami.5b03066] [PMID: 25899209]

[77]   Abbaspour, A.; Norouz-Sarvestani, F.; Noori, A.; Soltani, N. Aptamer-conjugated silver nanoparticles for electrochemical dual-aptamer-based sandwich detection of staphylococcus aureus. *Biosens. Bioelectron.*, **2015**, *68*, 149-155.
[http://dx.doi.org/10.1016/j.bios.2014.12.040] [PMID: 25562742]

[78]   Friedman, A.D.; Kim, D.; Liu, R. Highly stable aptamers selected from a 2′-fully modified fGmH RNA library for targeting biomaterials. *Biomaterials*, **2015**, *36*, 110-123.
[http://dx.doi.org/10.1016/j.biomaterials.2014.08.046] [PMID: 25443790]

[79]   Li, H.; Hu, H.; Zhao, Y.; Chen, X.; Li, W.; Qiang, W.; Xu, D. Multifunctional aptamer-silver conjugates as theragnostic agents for specific cancer cell therapy and fluorescence-enhanced cell imaging. *Anal. Chem.*, **2015**, *87*(7), 3736-3745.
[http://dx.doi.org/10.1021/ac504230j] [PMID: 25686206]

[80]   Mondal, S.; Verma, S. Catalytic and SERS Activities of Tryptophan-EDTA Capped Silver Nanoparticles. *Z. Anorg. Allg. Chem.*, **2014**, *640*(6), 1095-1101.
[http://dx.doi.org/10.1002/zaac.201400056]

[81]   Dojčilović, R.; Pajović, J.D.; Božanić, D.K.; Bogdanović, U.; Vodnik, V.V.; Dimitrijević-Branković, S.; Miljković, M.G.; Kaščaková, S.; Réfrégiers, M.; Djoković, V. Interaction of amino acid-functionalized silver nanoparticles and Candida albicans polymorphs: A deep-UV fluorescence imaging study. *Colloids Surf. B Biointerfaces*, **2017**, *155*, 341-348.
[http://dx.doi.org/10.1016/j.colsurfb.2017.04.044] [PMID: 28454063]

[82]   Chandra, A.; Singh, M. Biosynthesis of amino acid functionalized silver nanoparticles for potential catalytic and oxygen sensing applications. *Inorg. Chem. Front.*, **2018**, *5*(1), 233-257.
[http://dx.doi.org/10.1039/C7QI00569E]

[83]   de Matos, R.A.; Courrol, L.C. Biocompatible silver nanoparticles prepared with amino acids and a green method. *Amino Acids*, **2017**, *49*(2), 379-388.
[http://dx.doi.org/10.1007/s00726-016-2371-4] [PMID: 27896446]

[84]   Bonor, J.; Reddy, V.; Akkiraju, H.; Dhurjati, P.; Nohe, A. Synthesis and Characterization of L-Lysine Conjugated Silver Nanoparticles Smaller Than 10 nM. *Adv. Sci. Eng. Med.*, **2014**, *6*(9), 942-947.
[http://dx.doi.org/10.1166/asem.2014.1583] [PMID: 26478827]

[85]   Shankar, S.; Rhim, J-W. Amino acid mediated synthesis of silver nanoparticles and preparation of antimicrobial agar/silver nanoparticles composite films. *Carbohydr. Polym.*, **2015**, *130*, 353-363.
[http://dx.doi.org/10.1016/j.carbpol.2015.05.018] [PMID: 26076636]

[86]   Khan, M.M.; Kalathir, S.; Lee, J.; Cho, M.H. Synthesis of Cysteine Capped Silver Nanoparticles by Electrochemically Active Biofilm and their Antibacterial Activities. *Bull. Korean Chem. Soc.*, **2012**, *33*(8), 2592-2596.
[http://dx.doi.org/10.5012/bkcs.2012.33.8.2592]

[87]   Kasthuri, J.; Rajendiran, N. Functionalization of silver and gold nanoparticles using amino acid conjugated bile salts with tunable longitudinal plasmon resonance. *Colloids Surf. B Biointerfaces*, **2009**, *73*(2), 387-393.
[http://dx.doi.org/10.1016/j.colsurfb.2009.06.012] [PMID: 19577440]

[88]   Daima, H.K.; Selvakannan, P.R.; Kandjani, A.E.; Shukla, R.; Bhargava, S.K.; Bansal, V. Synergistic influence of polyoxometalate surface corona towards enhancing the antibacterial performance of tyrosine-capped Ag nanoparticles. *Nanoscale,* **2014**, *6*(2), 758-765.
[http://dx.doi.org/10.1039/C3NR03806H] [PMID: 24165753]

CHAPTER 4

Bioconjugation of Silver Nanoparticles

Abstract: The bioconjugation of AgNPs requires their colloidal stabilization in the green solvents. Preferably, the aqueous media provides a suitable environment for studying the biological properties of conjugated AgNPs and their physiological characteristics. Mainly, the conjugation of AgNPs with suitable moieties occurs *via* chemical conjugation such as sulfo-NHS coupling, EDC-coupling and by the approaches such as ligand addition, ligand exchange, and encapsulation. The bioconjugated AgNPs offer numerous applications in bioimaging, diagnostics, and molecular medicine. This chapter presents the bioconjugation strategies for AgNPs and their significance in the future biological applications.

Keywords: Bioconjugation, Diagnostics, EDC-coupling, Ligand addition, Ligand exchange, Sulfo-NHS coupling.

1. INTRODUCTION

Owing to their unique physical, optical, and electrical properties, the AgNPs, upon conjugation with biomolecules of interest, present interesting applications ranging from bioimaging to biodiagnostics [1, 2]. The conjugation of silver nanoparticles with biomacromolecules of interest occurs *via* three strategies, mainly ligand addition, ligand exchange and encapsulation [3]. Fig. (**1**) presents various surface functionalization strategies of AgNPs for their further conjugation with biomolecules. Mainly, the presence of a functional head group at the surface of AgNPs and the labile biomacromolecule facilitate the conjugation process [4]. The interactive forces that help stabilize the biomacromolecule-AgNPs complex include electrostatic interactions, hydration forces, van der Waals forces, and depletion stabilization [5]. The surface fabrication of AgNPs with organic ligands before their conjugation to biomacromolecules promotes their colloidal stabilization by preventing the agglomeration, in addition to providing the necessary auxiliary functionalities for conjugation to biomacromolecules [6, 7]. The generation of surface functionalized AgNPs occurring at elevated temperatures mostly in the organic solvents, results in an effective purging out of impurities [8]. The resultant crystalline structure of AgNPs appended with organic ligands achieves steric stabilization in the organic solvent [9, 10]. However, the organic phase stabilization of AgNPs fabricated with non-polar hydrophobic ligands and their subsequent destabilization in aqueous media limits their biological applications [11]. The destabilization causes a direct physiological exposure of the metallic Ag and Ag^+ ions, which proves detrimental to the biological components. Therefore, it prioritizes the presence of hydrophilic ligands as surface functionalization of AgNPs to ensure a reasonable aqueous compatibility for achieving optimal biological properties [12]. The presence of polar head functional groups such as $-NH_2$, -COOH, and –OH, in addition to –SH group with low polarity but higher binding affinity towards the AgNPs ensure a considerable hydrophilicity [13, 14]. As soon as the functionalized AgNPs appear in the aqueous phase, their biological applications and bioconjugation potency, gain precedence as majority of the organic solvents used to stabilize the colloidal AgNPs prove harmful and toxic towards the living tissues by causing direct necrosis, or indirect immunogenic manifestations [15, 16]. The aqueous stability of colloidal AgNPs provides a suitable medium for testing the potency of surface engineered nanoparticle towards various biological applications [17].

The conjugation of AgNPs with biomolecules following the various strategies listed in Fig. (1) requires the presence of a functional group prior to their tethering with the biomolecules of interest. As such, the direct conjugation of the biomolecules on AgNP scan also occur *via* various physical interactions, however, the strong covalent chemical conjugation of AgNPs with biomacromolecules occurs with functional head groups present on the surface decorated AgNPs by forming disulfide, amide, thiol, and amine bonds [18, 19]. The strong covalent conjugation between the biomolecules and surface functionalized AgNPs prevent the leaching out of nanoparticles, which may cause serious implications when present in a biological system [20, 21]. As such, the leached, naked AgNPs, when exposed directly to the tissues trigger immunogenicity, or may lead to oxidative stress due to the *in vivo* formation of Ag^+ ions [22 - 24]. Therefore, the nature of bonding between surface-functionalized AgNPs and biomolecule of interest presents high importance for the biological applications of AgNPs. The formulations based on AgNPs deliberated as antimicrobial agents release Ag^+ ions that cover with proteinaceous biofilms by the microbial cells, which reportedly prepare the microbial cells towards silver ion mediated stress [25 - 27]. Similarly, the vulnerability of microbial cells sometimes lowers due to this phenomenon due to genetic alterations caused by nanoparticles [28 - 30]. Nevertheless, the presence of non-functionalized AgNPs promotes microbial cell necrosis by a cascade of events instigated by their internalization in the microbial cells.

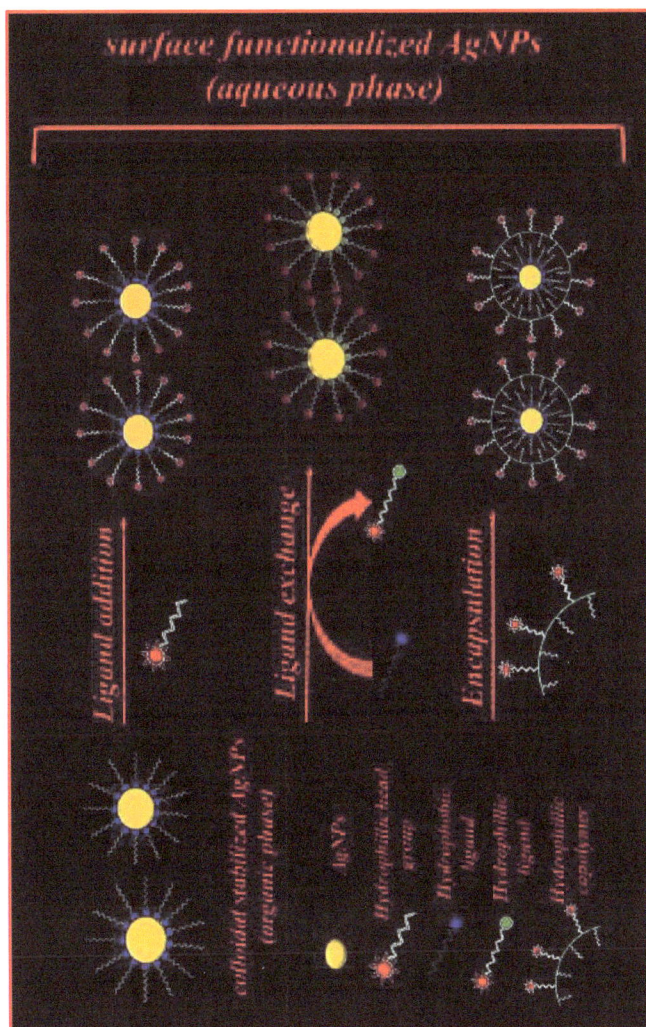

Fig. (1). Surface functionalization strategies of AgNPs for further bioconjugation.

2. STRATEGIES FOR BIOCONJUGATION

2.1. Ligand Addition

This strategy revolves around the addition of a functional group containing ligands directly on the surface of AgNPs for generating a functionalized nanosystem for further conjugation with the biomacromolecule of interest [31].

The addition of thiol ligand at the surface of AgNPs presents highly effective ligand addition strategy to promote the further bioconjugation of AgNPs, mainly with the enzymes for the immobilization, and inactivation of the latter [32, 33]. Other applications of thiol-functionalized AgNPs include their conjugation with antibodies for an effective delivery of the therapeutic genes [34]. The reduction of thiol-functionality on AgNPs surface further results in the formation of disulfide-functionalized nanoparticles that mainly form covalent linkages with the corresponding thiolated biomacromolecule [35]. This process requires a pre-modification of the AgNPs surface with molecules that serve as sources of thiol group. These include: 1,2-distearoyl-sn-glycero-3-phosphoethanola-ine-N-[PDP(polyethylene glycol)] (PDP-PEG-DSPE), 3-pentadecyl phenol–4-sulphonic acid (PDPSA), thioalkyl-PEG-R (R = H, CH_3, CH_2COOH, NH_2), alkylthiols: $HS(CH_2)_nCH_3$ (n=11), thioalkyl amines: $HS(CH_2)_nNH_2$ (n=2), thioalkyl acids: $HS(CH_2)_nCH_3$ (n = 10), or 1,2-dioleoyl-sn-glyce-o-3-phosphoethanolamine-N-[3-(2-pyridyldithio)propionate] (PDP-PE) [36]. The thiol donating groups accelerate the formation of the covalent bond between AgNPs and biomacromolecule. The electrostatic interactions, covalent or dative linkages between the AgNPs with metallic sulfur promote a higher stability of the conjugation between thiolated biomolecules and surface stabilized AgNPs. As such, the thiol-group containing biomolecules such as homocysteine, glutathione, cysteine, and cysteamine display high-binding energy non-covalent electrostatic interactions with the surface of AgNPs in acidic medium, hence serving as the linker molecules for anchoring with a biomolecule of interest.

Alternatively, the thiolated biomolecule conjugates with amine-functionalized AgNPs in the presence of coupling agents such as sulfo-SMCC (sulfosuccinimidyl-4-(maleimidomethyl) cyclohexane-1-carboxylate [37]. Similarly, the Michael addition between the aldehyde-functionalized AgNPs with amine containing biomolecule yields maleimide-tagged conjugates of AgNPs with biomolecules [38]. Notably, the thiolated AgNPs undergo a nucleophilic addition reaction on the maleimide-tagged biomolecule at room temperature in aqueous medium, to form thioether bond with C3/or C4 carbon atom [39]. However, the occurrence of several rearrangement side-reactions results in the formation of AgNP-disulfides, which hamper the selectivity of the reaction. However, the presence of activating reagents such as N-succinimidyl-3-(2-pyridyldithio) propionate (SPDP), or N-succinimidyl-S-acetylthioacetate (SATA) significantly improve the reaction efficiency [40, 41]. The maleimide tagged antibodies bioconjugated to AgNPs present applications as drug delivery vectors.

Carboxylates, citrate ions, and polyoxoanions present a stabilizing effect on the surface engineered AgNPs by directly interacting with the surface, hence providing an overall negative charge [42]. The negative charge ensures an electrostatic stabilization of the colloidal solution of carboxyl fabricated AgNPs. Alternately, the tethering of amino functionality or polyethyleneimine appendage provides a positive charge on the AgNPs surface, which offers colloidal stabilization by electrostatic interactions. Similarly, the functionalization of AgNPs surface with thiomalic acid [43], thioctic acid, mercaptoacetic acid, mercaptopropionic acid, mercaptohexanoic acid, 11-mercaptoundecanoic acid, and mercaptopropionic sulfonic acid present bifunctional interaction *via* 'S' atoms and free carboxylate group hence maintaining the hydrophilicity of surface functionalized AgNPs [44]. The ligands such as thiobarbiturates offer interactions *via* sulfur and nitrogen atoms with free carboxylic group available for conjugation with the desirable biomacromolecule [45], hence improving the aqueous dispersion of functionalized AgNPs. Long-chain carboxylates such as cis-1--eicosenoic acid, palmitoleic acid, linoleic acid, linolenic acid, and elaidic acid interact with AgNPs *via* olefenic group with carboxylate functionality not involved in interactions with AgNP surface [46 - 48]. The conjugation of amino/ carboxylate functionalized AgNPs occurs with desired biomolecule mainly *via* the formation of amide bond. The formation of bioconjugated amide bond involves activation of the carboxylic group on biomolecule by 1-ethyl-3-(3-dimethylaminopropyl) carbodiimide (EDC) to yield O-acylisourea ester intermediate, which further conjugates to amine functionality present on the AgNPs surface *via* amide bond [49 - 51]. This approach as well extends for the conjugation of amine functionalized AgNPs with phosphate groups on nucleic acids *via* the formation of N-hydrosuccinimide ester intermediate. EDC-mediated amide bond formation maintains the hydrophobicity of bioconjugated AgNPs and the biotolerance of the nanosystem by substituting the utilization of amide bond forming chemical reagents [52]. Moreover, the participating surface tagged AgNPs and biomolecules do not undergo chemical modifications that might lower their bioactivity. However, the stability of ester intermediate poses a significant challenge in bioconjugation, which improves in the presence of reagents N-hydroxysulfosuccinimide (sulfo-NHS), or N-hydroxysuccinimide (NHS) (Fig. **2**). Nevertheless, the EDC coupling approach causes self-polymerization of the participating biomolecules, hence requiring a periodical elimination of the excess of the reagent from the reaction mixture prior to its conjugation with the surface functionalized AgNPs [53 - 56]. The EDC coupling requires specific physicochemical conditions for amide bond formation that discourage the orientation immobilization of the biomolecules such as antibodies (Fig. **2**).

A. EDC coupling

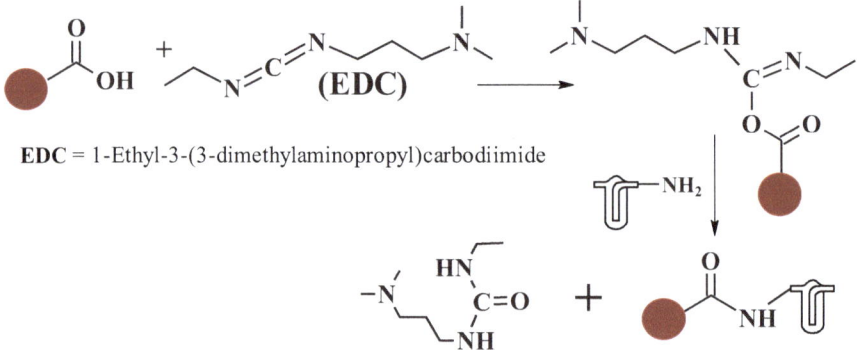

EDC = 1-Ethyl-3-(3-dimethylaminopropyl)carbodiimide

B. Sulfo-NHS coupling

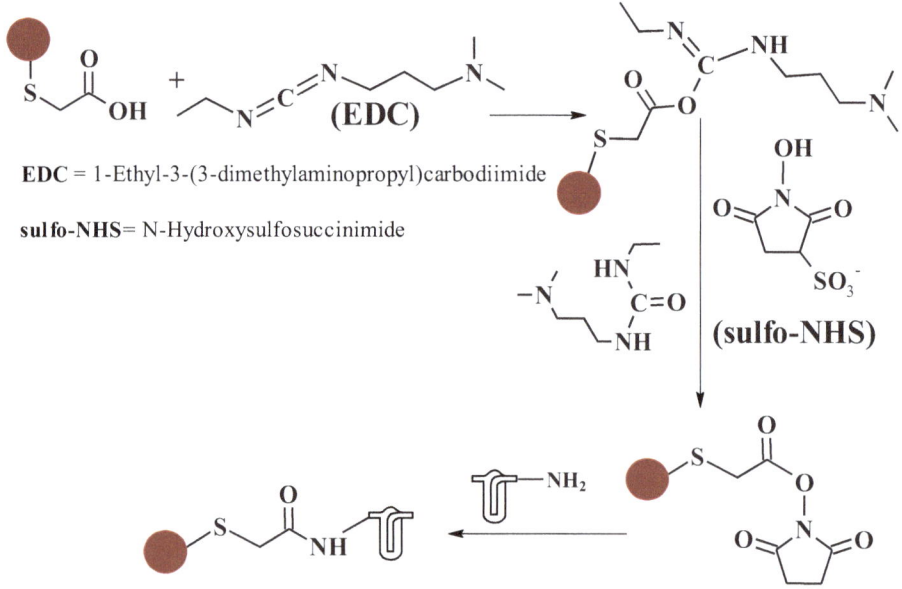

EDC = 1-Ethyl-3-(3-dimethylaminopropyl)carbodiimide

sulfo-NHS = N-Hydroxysulfosuccinimide

AgNPs   Biomolecule

Fig. (2). Chemical conjugation of AgNPs with biomolecules.

Pollok *et al.* (2019) reported the orientation controlled bioconjugation of antibodies to AgNPs by using heterobifunctional cross-linkers (HBCLs) *via* hydrazone conjugation approach. The thiol or dithiolane group anchors to the AgNPs, *via* alkyl chain or PEGylation, which immobilized the antibody immunoglobulin G on the nanoparticle surface. The half-metalloimmunoassay suggested that the antibody conjugated AgNPs *via* heterobifunctional cross-linkers displayed higher activity and prolonged stability compared to the physisorption of antibodies on AgNPs [20]. These investigations indicated that the heterobifunctional cross-linkers apply orientational control over the antibodies. Thomas *et al.* (2019) reported the binding of chlorogenic acid capped AgNPs with ctDNA mainly *via* hydrogen bonding, electrostatic interaction and van der Waals forces. Here, the chlorogenic acid acts as both the capping and reducing agent for AgNPs in order to achieve enthalpy driven, strong exothermic binding while complexation with ctDNA [57]. Zhang *et al.* (2012) reported highly stabilized pH dependent conjugation of monothiolated DNA to AgNPs. The optimal bioconjugation of DNA occurred at acidic pH, while decelerated at pH 5. The AgNP-DNA conjugates contained high density of DNA molecules that maintain their function for the recognition of complementary DNA. These findings proved highly beneficial for AgNPs conjugation to DNA at low pH, which usually causes the loss of the charge density on nanoparticle and causes the protonation of adenine and cytosine bases of DNA [58]. Yu *et al.* (2016) reported the binding strength of thiolated nucleosides and nucleobases conjugated to AgNPs by colorimetric analysis. The analysis confirmed that the AgNPs interact with thiolated oligonucleotides *via* robust, sequence-dependent interactions owing to the presence of 'N' atoms and –C=O group on pyrimidine and purine rings. The click chemistry approach presents another useful strategy for the generation of bioconjugated AgNPs as it frequently yields easily purifiable products in high yield in the presence of mild solvents. This approach involves the conjugation of AgNPs with biomolecules *via* highly stable triazole linker, which acts as amide bond bioisostere, while displaying an extraordinary stability against the enzymatic degradation. The marked biocompatibility of azoles validates the application of click chemistry in the bioconjugation of AgNPs with biomolecules. Besides these advantages, the click reactions require the synthesis of azido/ alkyne reactants prior to the triazole cycloaddition, which lowers the product yield and reactant efficacy, thereby making it a less profitable strategy for the bioconjugation of AgNPs [59]. Fig. (**3**) represents the various conjugation strategies for AgNPs to interact with biomolecules.

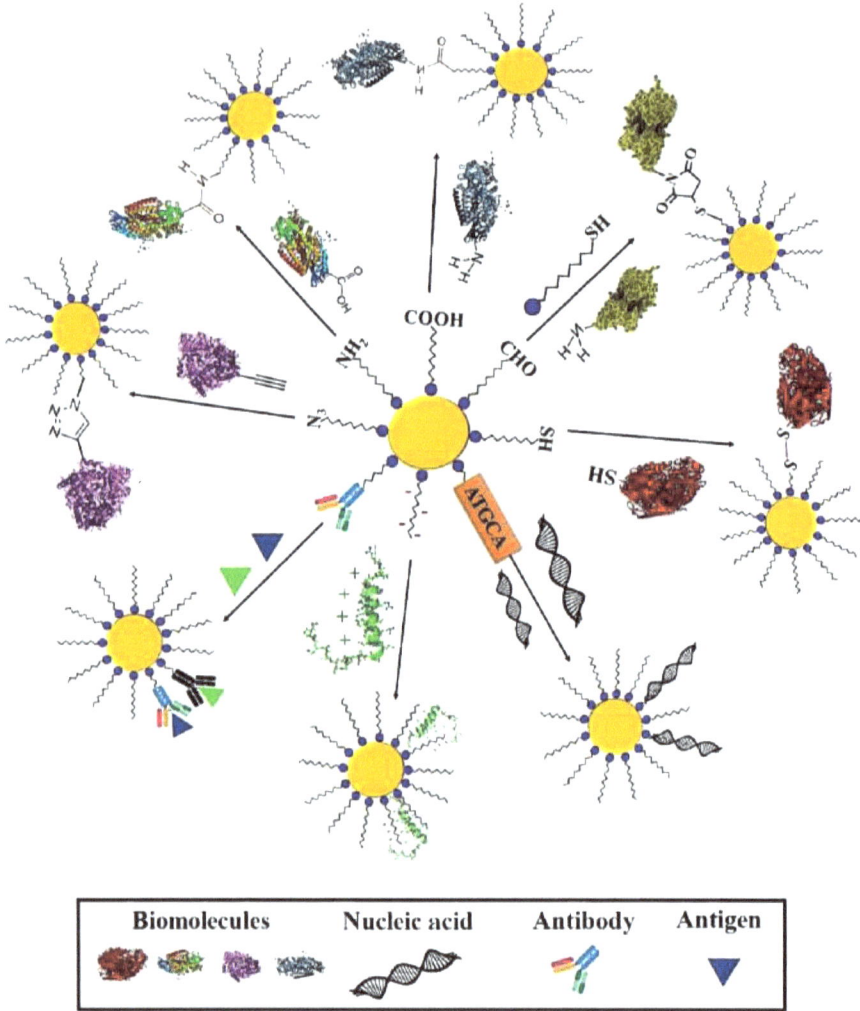

Fig. (3). Conjugation strategies for AgNPs.

2.2. Ligand Exchange

The methodology of ligand exchange encompasses a replacement of the hydrophobic chains appended hydrophobic chains on the AgNPs surface with a suitable hydrophilic ligand in order to achieve colloidal stability in the aqueous phase. The ligand exchange occurs by incorporating the AgNPs with $-NH_2$, $-PH_3$,

-SH, and –COOH functionalities, that promote the cellular infiltration of functionalized AgNPs, and provide electrostatic stabilization in response to adverse physicochemical conditions. Similarly, the functionalities such as polyethylene glycol involve hydrogen bonding interactions that stabilize the AgNPs in the aqueous medium. AbdulHalim *et al.* (2014) reported a novel approach for the thiolated ligand exchange on AgNPs, while preserving their characteristic properties. The AgNP films obtained by this method presented a smooth surface morphology that offered versatile applications. This method enabled the replacement of native ligands anchored to AgNPs with new ligands containing a variety of functional groups to obtain smooth thin films that presented applications in solution processed devices [60]. Jaskolska *et al.* (2019) reported competition-driven exchange of ligands for the surface functionalization of AgNPs and AgNP nanoclusters with a superior colloidal stability. The competition-driven ligand exchange utilizes a substrate that possesses a high affinity towards the primary ligand, thereby activating the colloidal dispersed AgNPs leading to their functionalization with a secondary ligand. Importantly, the secondary ligand does not participate in competition with the substrate and undergoes no exchange in the absence of substrate [61]. Lopez-Lorente *et al.* (2014) reported the capillary electrophoretic behavior of AgNPs coated with citrate in aqueous media for performing the ligand exchange reactions with thiol. The selective functionalization of citrate capped AgNPs occurs in capillary on the addition of thiol compounds such as thiomalic acid and thioctic acid in electrophoretic buffer leading to the formation of Ag-S bonds that displayed self-assembled monolayers with superior orientation and high order, compared to the citrate ligand [62]. Sardar *et al.* (2007) reported the synthesis of AgNPs in the presence of poly(allylamine) that acts as a reducing and stabilizing agent in aqueous media. The obtained polymer-functionalized AgNPs displayed marked stability in water at room temperature for one month. The ligand exchange on the polymer-functionalized AgNPs occurred rapidly with ω-functional –OH, -COOH, and $–NH_2$ carrying alkylthiols to yield finely dispersed nanoparticles [63]. Ko *et al.* (2013) developed an advanced method for synthesizing nanocomposite multilayers to achieve the densely packed assembly of hydrophobic AgNPs decorated with a wide range of functional groups. The colloidal AgNPs in an organic solvent undergo layer-by-layer assembly prompted by $–NH_2$ fabricated polymers, hence forming the nanocomposite multilayers. These multilayer films formed on colloidal AgNP substrate promote the adsorption of electrostatically charged layers by foregoing the surface treatments [64]. Lv *et al.* (2017) reported the electrochemical detection of glutathione intracellularly, based on ligand exchange strategy for releasing DNA-decorated AgNPs. The synthesis of AgNPs occurred by reducing the Ag^+ ions on DNA template, which releases the AgNPs through highly stable Ag-S interactions with glutathione. The electrochemical response appeared when glutathione-fabricated AgNPs released Ag^+ ions causing an electrochemical response. This approach effectively detected intracellular glutathione in the range $0.1 - 1$ µM, with high sensitivity at a detection limit 2.3 x 10^{-11} M. COX *et al.* (2020) developed metallic nanoclusters with marked aqueous solubility. Unlike most water-soluble nanoclusters that require passivating ligands for stabilization, the reported nanosystem undergoes ligand induced structural alterations as detected in resistive-pulse nanopore sensing. The nanopore system promoted time-resolved appraisal of ligand addition, and their exchange. As such, the exchanging of thiolated poly(ethylene glycol) ligands with glutathione fabricated metal nanoparticles, and tiopronin displayed a rapid rate of exchange in the nanoconfined area of the pore. This phenomenon offered a rapid sensing of peptides [65]. Cho *et al.* (2018) reported propanethiol-ligand exchanged AgNPs with superior thermal stability for improved dispersion in perovskite solar cells. The AgNPs capped with polyvinyl propylene aggregate in organic media, but their aggregation reduced on exchanging the ligand with propanethiol. The propanethiol capped AgNPs displayed improved thermal stability, and ameliorated light absorption by promoting scattering. This phenomenon enhanced the charge generation that provided applications in solar cells [66]. Lee *et al.* (2017) developed AgNPs based thin film electrodes by exchanging oleylamine ligand with acrylic acid. The lowering of sintering temperature for the AgNPs presented applications in polymeric devices and electronics. The thin films made of AgNPs provided conductivity of only one order of magnitude lower than the bulk Ag on sintering at 150 °C. Importantly, the reported methodology provided applications in flexible electronics by bypassing the vacuum requirement and high-temperature processes. The work presented the optimal extent of exchange of ligands which prevented cracking caused by volume shrinkage of the capped ligands [67]. Merg *et al.* (2017) demonstrated the surface tuning of nanoparticle surface with ligand exchange approach. The peptide-based assemblies of nanoparticles serve a fitting candidature for ligand exchange process that effectively stabilize and functionalize the superstructures of nanoparticles, hence preventing their aggregation in colloidal solution. The peptide ligands that functionalize the surface of hollow or spherical AgNP superstructures exchange effectively with thiolated ligands whilst preserving the stability of the basic structure. These properties ensure a successful disposition of the nanoparticle superstructures for their target applications [68]. Long *et al.* (2017) prepared three types of AgNPs by ligand exchange method from the precursor citrate-capped AgNPs, where (1R)-menthyl hexyl phosphonate, 6-mercapto hexanoic acid, and mercaptopropane sulfonate ligands exchange with citrate. The resultant nanoparticle provided substantial antibacterial activity against *Escherichia coli* by causing oxidative stress, lipid peroxidation, and membrane rupture in the microbial cells. The citrate linker exhibits weaker binding affinity towards AgNP core, which allows the absorption exchange with atmospheric dioxygen, thereby causing oxidative dissolution of Ag^0. Similarly, the terminal sulfonate group in MPS displays lower capacity to release silver ions compared to carboxyl group that results in lesser dissolution and more release of ionic silver retaining on the surface of AgNPs [69]. Suzuki *et al.* (2019) developed a ligand exchange reaction for the highly sensitive detection of dopamine in serum. The system consisted of a fluorescent probe with boron-dipyrromethenyl ligand as the main fluorophore attached to the metal nanoparticles. The nanosystems displayed heightened fluorescence on interacting with dopamine, due to the release of the central metal ions, hence indicating their interactions with dopamine. The nanosystems efficiently detected dopamine at 1.1 nM concentration, with no pronounced effect caused by the presence of foreign substances [70].

2.3. Encapsulation

Encapsulation caters to the phase transferring of surface decorated AgNPs from organic media to aqueous medium by coating the original hydrophobic ligand with amphiphilic ligands, so that the hydrophilic portion of amphiphilic ligand points outwards to the parent solution. The hydrophilic surface ligand containing polar groups offers an enhanced stability in an aqueous solution, thereby promoting their further conjugation with biomolecules for enhanced applications. Song *et al.* (2019) performed the encapsulation of AgNPs with hydrogels constituted of zwitter ions for developing highly efficient antifouling catalysis. The polycarboxybetaine-AgNPs effectively removed the priority pollutants such as 4-nitrophenol by reducing it to 4-aminophenol, reaching >95% conversion efficiency within 5 minutes [71]. Qasim *et al.* (2018) reported poly--isopropylacrylamide-based polymeric AgNPs with marked antimicrobial properties against the gram-positive and gram-negative bacteria mainly by bacteriostatic mode of action. The AgNPs are completely encapsulated inside the core of the AgNP-polymeric nanoparticle due to a sturdy localization and stabilization offered by the characteristic 3-D network structure presented by nanosystem. In addition, the presence of functional head-groups –CONH-, -OH, and –COOH promoted surface adsorption of AgNPs, thereby leading to their stabilization [72]. Zaheer *et al.* (2016) reported the reversible encapsulation of AgNPs into the helical structure of amylose, with the assistance of diosgenin as a reducing agent. The analysis by transmission electron microscopy indicated the formation of diverse shapes, including rod-shaped, triangular, spherical, polydispersed, hexagonal, and aggregated AgNPs. The presence of –OH groups on amylose readily promoted the adsorption on positively charged AgNPs. The biodegradable nature of amylose presented drug delivery applications of the reported nanosystem, with a controlled release profile for improved drug pharmacokinetics [73]. Khan *et al.* (2019) reported the reversible encapsulation of AgNPs into amylose helix resulting in the formation of chain-like structure. The nanosystem reportedly offered antimicrobial properties against *E. coli*, and *S. aureus*. Mainly, the metal binding tendency of cysteine caused a reduction of adsorbed silver ions on AGNPs surface present in the protein wall. The presence of AgNPs notably enhanced the decomposition temperature of amylose, which also depended on the content of hydrogen peroxide required for the carboxyl and carbonyl degree of oxidation of encapsulating amylose [74]. Sathiyaseelan *et al.* (2020) reported chitosan encapsulated phytogenic AgNPs with enhanced biological applications owing to the biocompatibility of the fungal chitosan. The polydispersed AgNPs displayed an average size <100 nm, with inhibition profile against several gram-positive and gran-negative bacterial strains with IC50 in the range 4.08-8.25 µg/ mL. The bioactivity mainly arises due to the presence of phenolic groups and flavonoids on the surface of phytogenic AgNPs [75]. Liu *et al.* (2014) reported dendrimer encapsulation of AgNPs with marked anti-inflammatory activities mainly due to the suppression of TNF-α, and IL-6 *in vitro* by 36.6% and 33.6%, respectively. The nanosystem proved highly effective in curbing the lipopolysaccharide-induced inflammation, while the AgNP-dendrimer complex displayed a synergistic effect among the treated animal models that displayed a faster rate of healing compared to the animals treated individually with AgNPs of dendrimer. Further investigations suggested that the N-atoms present in polymeric backbone of dendrimer enabled coordination and stabilization of AgNPs, thereby extending their applications in drug-delivery [76]. Silva *et al.* (2015) reported bactericidal properties of doxycycline conjugated polyvinylpyrrolidone-encapsulated AgNPs against *E. coli*. The polyvinylpyrrolidone polymer stabilized AgNPs and prevented their excessive growth and aggregation. Doxycycline directly bounded to the polyvinylpyrrolidone, which caused an enhancement in its concentration around AgNPs, thereby potentiating the bactericidal effect of the drug. The culmination in the antimicrobial response by the doxycycline carrying nanosystem presented applications in targeting the multidrug resistant microbes, where the efflux pumps discourage the maintenance of an effective drug concentration for an optimal effect [77]. Ghodake *et al.* (2019) reported peptide encapsulated AgNPs for the highly sensitive, colorimetric detection of Pd(II), caused mainly by the formation of coordination complex between the peptide ligands and Pd(II). The interaction cause aggregation of AgNPs, thereby causing colorimetric changes in the colloidal solution. Interestingly, the nanoprobes imparted a swift detection of the Pd(II) ions, without the interference from the other ions present in the solution hence offering applications for the detection of palladium contamination. Mainly, nanoprobes consisting of the whey-protein encapsulated AgNPs employed here offered analytical utility for the detection of trace amount of Pd(II) appearing during the catalytic reactions [78]. Taheri *et al.* (2015) reported phospholipid encapsulated AgNPs conjugated to antibacterial plasma polymer films for obtaining antimicrobial coatings. These antimicrobial coatings displayed marked inhibition of gram-positive and gram-negative bacteria, including *S. aureus, P. aeruginosa*, and *S. epidermidis*. The coatings reportedly demonstrated negligible cytotoxicity and trivial immunogenicity against human dermal fibroblasts, and bone marrow derived macrophages, respectively. These coatings showed marked inhibition of bacterial growth and lowered the expression of inflammation causing cytokines from the bone marrow derived macrophages, thereby indicating a decreased inflammatory response. These properties indicated the application of antibacterial plasma polymer films conjugated with phospholipid encapsulated silver nanoparticles as wound dressings [79]. Qasim *et al.* (2018) reported antimicrobial potency of poly-N-isopropylacrylamide encapsulated AgNPs towards gram-positive and gram-negative bacteria, where the bioactivity depended on the amount and size of nanoparticles in the polymeric matrix. The nanosystem demonstrated bacteriostatic activity pertaining to the postsurgical infections of biomedical implants. The nanosystem containing large sized AgNPs proved physiologically benevolent towards human cell lines, compared to the nanoparticles with finer size, owing to the lower ability to penetrate the biological membranes [72]. Fujiwara *et al.* (2017) reported AgNPs encapsulated in hollow silica spheres for an efficient removal of sulfur-containing compounds. The nanosystem acting as adsorbent material consisted of hollow silica spheres containing propyl amine groups anchored inside the silica shells. The deposition of AgNPs inside these silica shells occurred by microwave irradiation. Interestingly, the nanosystem demonstrated adsorption of low concentration of volatile sulfur compounds including H_2S, and *tert*-butylmercaptan at ambient conditions, despite adsorbing protein molecules mainly due to the deactivation of AgNPs by the latter. The nanosystem exhibited marked reusability, while retaining its original adsorption potency even after successive adsorption cycles [80]. Pavoski *et al.* (2019) developed silica encapsulated AgNPs as antibacterial fillers in ethylene polymerization. The bactericidal action appeared against *S. aureus* and *E. coli* mainly due to the interruption in bacterial proliferation, hence offering application as active medicine packaging material as the silica prevents the escape of encapsulated AgNPs. The proposed nanomaterial presented applications as antibacterial thermoplastics with packaging properties of polyethylene [81]. Karunamuni *et al.* (2016) reported the properties of silica-encapsulated AgNPs as contrast agents for dual-energy mammography. The nanoparticles of average size 102 nm exhibited half-life of 15 minutes in the systemic circulation and eventually accumulated in the macrophage-rich organs, including spleen, liver, and lymph nodes, hence acting as effective contrast agent for dual-energy x-ray imaging [82].

Bioconjugation of AgNPs offers ubiquitous applications in various disciplines that led to the advancements in the medical field. From image-enhancing agents to gene therapy, and from antibody recognition to molecular medicine, Fig. (**4**) illustrates the various applications of bioconjugated AgNPs.

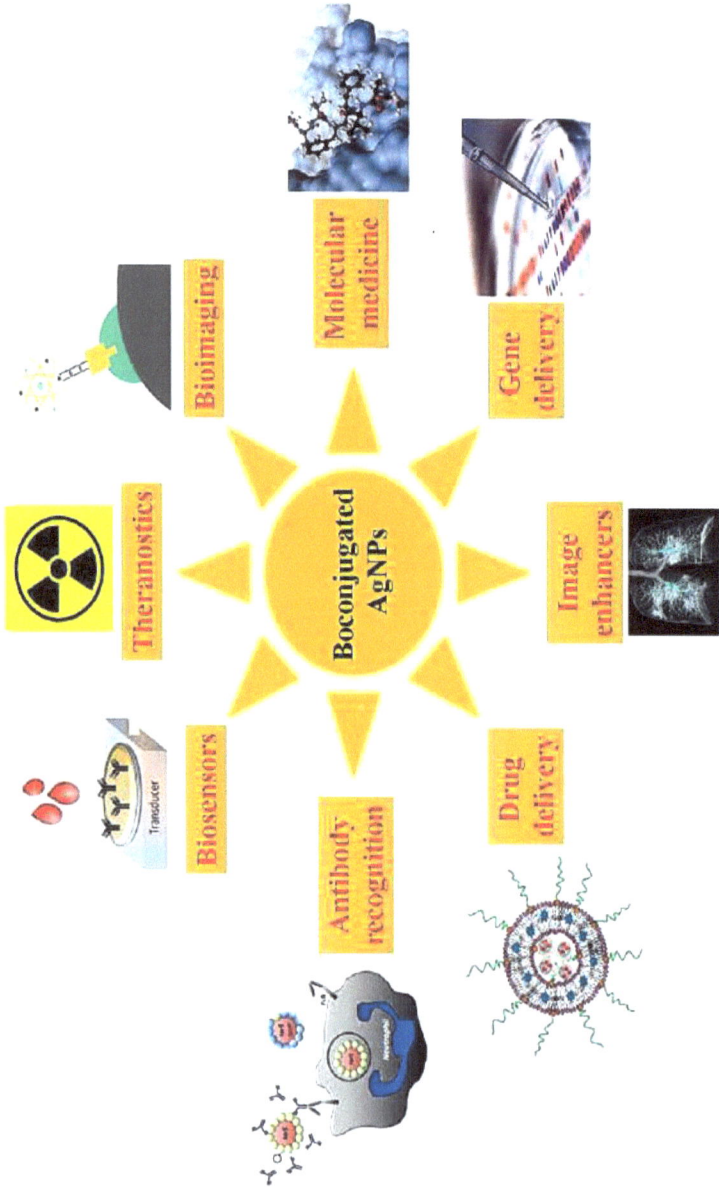

Fig. (4). Various applications of bioconjugated AgNPs.

CONCLUSION

The bioconjugation of AgNPs presents advanced applications in molecular medicine, drug delivery, and tomography image enhancement. The surface fabrication of AgNPs with suitable functionalities and their further tethering to the therapeutics or biomolecule of interest *via* covalent bonding or electrostatic interactions minimizes the nanoparticle toxicity and improves their physiological tolerance for further applications. The AgNPs conjugate with antigens, antibodies, enzymes, peptides, and biotin *via* EDC coupling, maleimide coupling or sulfo-NHS coupling to achieve biological applications. However, as soon as the bioconjugated AgNPs enter the cells, they may lose their functionalities hence posing toxicity to the parent cells. Therefore, the application of bioconjugated AgNPs must take into consideration the imminent toxicity to the host cells for clinical success.

REFERENCES

[1]   Riaz Ahmed, K.B.; Nagy, A.M.; Brown, R.P.; Zhang, Q.; Malghan, S.G.; Goering, P.L. Silver nanoparticles: Significance of physicochemical properties and assay interference on the interpretation of *in vitro* cytotoxicity studies. *Toxicol. In Vitro,* **2017**, *38*, 179-192.
[http://dx.doi.org/10.1016/j.tiv.2016.10.012] [PMID: 27816503]

[2]   Fahmy, H.M.; Mosleh, A.M.; Elghany, A.A.; Eldin, E.S.; Serea, E.S.A.; Ali, S.A.; Shalan, A.E. Coated silver nanoparticles: synthesis, cytotoxicity, and optical properties. *RSC Advances,* **2019**, *9*(35), 20118-20136.
[http://dx.doi.org/10.1039/C9RA02907A]

[3]   Prasher, P.; Sharam, M.; Mudila, H.; Gupta, G.; Sharma, A.K.; Kumar, D.; Bakshi, H.A.; Negi, P.; Kapoor, D.N.; Chellappan, D.K.; Tambuwala, M.M.; Dua, K. Emerging trends in clinical implications of bio-conjugated silver nanoparticles in drug delivery. *Colloid Interface Sci. Commun.,* **2020**, *35*: 100244.
[http://dx.doi.org/10.1016/j.colcom.2020.100244]

[4]   Borowik, A.; Butowska, K.; Konkel, A.; Banasiuk, R.; Derewonko, N.; Wyrzykowski, D.; Davydenko, M.; Cherepanov, V.; Styopkin, V.; Prylutskyy, Y.; Pohl, P.; Krolicka, A.; Piosik, J. The Impact of Surface Functionalization on the Biophysical Properties of Silver Nanoparticles. *Nanomater. (MDPI),* **2019**, *9*, Article ID. 973.

[5]   Yang, S. Physical and Chemical Modification of Silver Nano Particles.*Polymer Nanocomposites Based on Silver Nanoparticles. Engineering Materials*; Lal, H.M.; Thomas, S.; Li, T.; Maria, H.J., Eds.; Springer: Cham, **2021**.
[http://dx.doi.org/10.1007/978-3-030-44259-0_3]

[6]   Ravindran, A.; Chandran, P.; Khan, S.S. Biofunctionalized silver nanoparticles: advances and prospects. *Colloids Surf. B Biointerfaces,* **2013**, *105*, 342-352.
[http://dx.doi.org/10.1016/j.colsurfb.2012.07.036] [PMID: 23411404]

[7]   Hussain, S.; Al-Thabaiti, S.A.; Khan, Z. Surfactant-assisted bio-conjugated synthesis of silver nanoparticles (AgNPs). *Bioprocess Biosyst. Eng.,* **2014**, *37*(9), 1727-1735.
[http://dx.doi.org/10.1007/s00449-014-1145-1] [PMID: 24556976]

[8]   Lee, S.H.; Jun, B-H. Silver nanoparticles: Synthesis and application for nanomedicine. *Int. J. Mol. Sci.,* **2019**, *20*(4), 865.
[http://dx.doi.org/10.3390/ijms20040865] [PMID: 30781560]

[9]   El-Nour, K.M.M.A.; Eftaiha, A.; Al-Warthan, A.; Ammar, R.A.A. Synthesis and application of AgNPs. *Arab. J. Chem.,* **2010**, *3*, 135-140.
[http://dx.doi.org/10.1016/j.arabjc.2010.04.008]

[10]   Yusuf, M. Silver Nanoparticles: Synthesis and Applications.*Handbook of Ecomaterials*; Martínez, L.; Kharissova, O.; Kharisov, B., Eds.; Springer: Cham, **2019**.
[http://dx.doi.org/10.1007/978-3-319-68255-6_16]

[11]   Zhang, Z.; Shen, W.; Xue, J.; Liu, Y.; Liu, Y.; Yan, P.; Liu, J.; Tang, J. Recent advances in synthetic methods and applications of silver nanostructures. *Nanoscale Res. Lett.,* **2018**, *13*(1), 54.
[http://dx.doi.org/10.1186/s11671-018-2450-4] [PMID: 29457198]

[12]   Pokhrel, L.R.; Dubey, B.; Scheuerman, P.R. Impacts of select organic ligands on the colloidal stability, dissolution dynamics, and toxicity of silver nanoparticles. *Environ. Sci. Technol.,* **2013**, *47*(22), 12877-12885.
[http://dx.doi.org/10.1021/es403462j] [PMID: 24144348]

[13]   Sooresh, A.; Kwon, H.; Taylor, R.; Pietrantonio, P.; Pine, M.; Sayes, C.M. Surface functionalization of silver nanoparticles: novel applications for insect vector control. *ACS Appl. Mater. Interfaces,* **2011**, *3*(10), 3779-3787.
[http://dx.doi.org/10.1021/am201167v] [PMID: 21957003]

[14]   Sanità, G.; Carrese, B.; Lamberti, A. Nanoparticle Surface Functionalization: How to Improve Biocompatibility and Cellular Internalization. *Front. Mol. Biosci.,* **2020**, *7*: 587012.
[http://dx.doi.org/10.3389/fmolb.2020.587012] [PMID: 33324678]

[15]   Ferdous, Z.; Nemmar, A. Health Impact of Silver Nanoparticles: A Review of the Biodistribution and Toxicity Following Various Routes of Exposure. *Int. J. Mol. Sci.,* **2020**, *21*(7), 2375.
[http://dx.doi.org/10.3390/ijms21072375] [PMID: 32235542]

[16]   Chakraborty, B.; Pal, R.; Ali, M.; Singh, L.M.; Shahidur Rahman, D.; Kumar Ghosh, S.; Sengupta, M. Immunomodulatory properties of silver nanoparticles contribute to anticancer strategy for murine fibrosarcoma. *Cell. Mol. Immunol.,* **2016**, *13*(2), 191-205.
[http://dx.doi.org/10.1038/cmi.2015.05] [PMID: 25938978]

[17]   Bayram, S.; Zahr, O.K.; Blum, A.S. Short ligands offer long-term water stability and plasmon tunability for silver nanoparticles. *RSC Advances,* **2015**, *5*(9), 6553-6559.
[http://dx.doi.org/10.1039/C4RA09667C]

[18]   Auria-Soro, C.; Nesma, T.; Velasco-Juanes, P.; Vinuela, A.L.; Gomez, H.F.; Fernandez, V.A.; Gongora, R.; Parra, M.J.A.; Roman, R.M.; Fuentes, M. Interactions of Nanoparticles and Biosystems: Microenvironment of Nanoparticles and Biomolecules in Nanomedicine. *Nanomater. (MDPI),* **2019**, *9*, Article ID. 1365.

[19]   Ahumada, M.; Suuronen, E.J.; Alarcon, E.I. Biomolecule Silver Nanoparticle-Based Materials for Biomedical Applications. *Handbook of Ecomaterials*; Martínez, L.; Kharissova, O.; Kharisov, B., Eds.; Springer: Cham, **2019**.
[http://dx.doi.org/10.1007/978-3-319-68255-6_161]

[20]   Pollok, N.E.; Rabin, C.; Smith, L.; Crooks, R.M. Orientation-Controlled Bioconjugation of Antibodies to Silver Nanoparticles. *Bioconjug. Chem.,* **2019**, *30*(12), 3078-3086.
[http://dx.doi.org/10.1021/acs.bioconjchem.9b00737] [PMID: 31730333]

[21]   Oh, J-H.; Park, D.H.; Joo, J.H.; Lee, J-S. Recent advances in chemical functionalization of nanoparticles with biomolecules for analytical applications. *Anal. Bioanal. Chem.,* **2015**, *407*(29), 8627-8645.
[http://dx.doi.org/10.1007/s00216-015-8981-y] [PMID: 26329278]

[22]   Flores-López, L.Z.; Espinoza-Gómez, H.; Somanathan, R. Silver nanoparticles: Electron transfer, reactive oxygen species, oxidative stress, beneficial and toxicological effects. Mini review. *J. Appl. Toxicol.,* **2019**, *39*(1), 16-26.

[http://dx.doi.org/10.1002/jat.3654] [PMID: 29943411]

[23] Homaee, M.B.; Ehsanpour, A.A. Silver nanoparticles and silver ions: Oxidative stress responses and toxicity in potato (Solanum tuberosum L) grown *in vitro. Hortic. Environ. Biotechnol.,* **2016**, *57*(6), 544-553.
[http://dx.doi.org/10.1007/s13580-016-0083-z]

[24] Patlolla, A.K.; Hackett, D.; Tchounwou, P.B. Silver nanoparticle-induced oxidative stress-dependent toxicity in Sprague-Dawley rats. *Mol. Cell. Biochem.,* **2015**, *399*(1-2), 257-268.
[http://dx.doi.org/10.1007/s11010-014-2252-7] [PMID: 25355157]

[25] Dakal, T.C.; Kumar, A.; Majumdar, R.S.; Yadav, V. Mechanistic Basis of Antimicrobial Actions of Silver Nanoparticles. *Front. Microbiol.,* **2016**, *7*, 1831.
[http://dx.doi.org/10.3389/fmicb.2016.01831] [PMID: 27899918]

[26] Yan, X.; He, B.; Liu, L.; Qu, G.; Shi, J.; Hu, L.; Jiang, G. Antibacterial mechanism of silver nanoparticles in *Pseudomonas aeruginosa*: proteomics approach. *Metallomics,* **2018**, *10*(4), 557-564.
[http://dx.doi.org/10.1039/C7MT00328E] [PMID: 29637212]

[27] Riaz, M.; Mutreja, V.; Sareen, S.; Ahmad, B.; Faheem, M.; Zahid, N.; Jabbour, G.; Park, J. Exceptional antibacterial and cytotoxic potency of monodisperse greener AgNPs prepared under optimized pH and temperature. *Sci. Rep.,* **2021**, *11*(1), 2866.
[http://dx.doi.org/10.1038/s41598-021-82555-z] [PMID: 33536517]

[28] Garraus, A-R.; Azqueta, A.; Vettorazzi, A.; Cerain, A.L. Genotoxicity of silver nanoparticles. *Nanomater. (MDPI),* **2020**, *10*, Article ID. 251.

[29] Tymoszuk, A.; Kulus, D. Silver nanoparticles induce genetic, biochemical, and phenotype variation in chrysanthemum. *Plant Cell Tissue Organ Cult.,* **2020**, *143*(2), 331-344.
[http://dx.doi.org/10.1007/s11240-020-01920-4]

[30] Butler, K.S.; Peeler, D.J.; Casey, B.J.; Dair, B.J.; Elespuru, R.K. Silver nanoparticles: correlating nanoparticle size and cellular uptake with genotoxicity. *Mutagenesis,* **2015**, *30*(4), 577-591.
[http://dx.doi.org/10.1093/mutage/gev020] [PMID: 25964273]

[31] Heuer-Jungemann, A.; Feliu, N.; Bakaimi, I.; Hamaly, M.; Alkilany, A.; Chakraborty, I.; Masood, A.; Casula, M.F.; Kostopoulou, A.; Oh, E.; Susumu, K.; Stewart, M.H.; Medintz, I.L.; Stratakis, E.; Parak, W.J.; Kanaras, A.G. The role of ligands in the chemical synthesis and applications of Inorganic nanoparticles. *Chem. Rev.,* **2019**, *119*(8), 4819-4880.
[http://dx.doi.org/10.1021/acs.chemrev.8b00733] [PMID: 30920815]

[32] Jiang, H.S.; Zhang, Y.; Lu, Z.W.; Lebrun, R.; Gontero, B.; Li, W. Interaction between silver nanoparticles and two dehydrogenases: Role of thiol groups. *Small,* **2019**, *15*(27): e1900860.
[http://dx.doi.org/10.1002/smll.201900860] [PMID: 31111667]

[33] Suzuki, H.; Chiba, H.; Futamata, M. Efficient immobilization of silver nanoparticles on metal substrates through various thiol molecules to utilize a gap mode in surface enhanced Raman scattering. *Vib. Spectrosc.,* **2014**, *72*, 105-110.
[http://dx.doi.org/10.1016/j.vibspec.2014.03.002]

[34] Arruebo, M.; Valladares, M.; Fernandez, A.G. Antibody-conjugated nanoparticles for biomedical applications. *J. Nanomater.,* **2009**, *2009*: 439389.
[http://dx.doi.org/10.1155/2009/439389]

[35] Lin, X.; O'Reilly Beringhs, A.; Lu, X. Applications of Nanoparticle-Antibody Conjugates in Immunoassays and Tumor Imaging. *AAPS J.,* **2021**, *23*(2), 43.
[http://dx.doi.org/10.1208/s12248-021-00561-5] [PMID: 33718979]

[36] Vargas, K.M.; Shon, Y-S. Hybrid lipid-nanoparticle complexes for biomedical applications. *J. Mater. Chem. B Mater. Biol. Med.,* **2019**, *7*(5), 695-708.
[http://dx.doi.org/10.1039/C8TB03084G] [PMID: 30740226]

[37] Chen, J.; Guo, L.; Qiu, B.; Lin, Z.; Wang, T. Application of ordered nanoparticle self-assemblies in

surface-enhanced spectroscopy. *Mater. Chem. Front.,* **2018**, *2*(5), 835-860.
[http://dx.doi.org/10.1039/C7QM00557A]

[38]   Hartlen, K.D.; Ismaili, H.; Zhu, J.; Workentin, M.S. Michael addition reactions for the modification of gold nanoparticles facilitated by hyperbaric conditions. *Langmuir,* **2012**, *28*(1), 864-871.
[http://dx.doi.org/10.1021/la203662n] [PMID: 22085199]

[39]   Baldwin, A.D.; Kiick, K.L. Tunable degradation of maleimide-thiol adducts in reducing environments. *Bioconjug. Chem.,* **2011**, *22*(10), 1946-1953.
[http://dx.doi.org/10.1021/bc200148v] [PMID: 21863904]

[40]   Obzansky, D.M.; Tseng, S.Y. Situ process for production of conjugates. *US5324650A,* **1994**.

[41]   Bae, E.; Park, H-J.; Yoon, J.; Kim, Y.; Choi, K.; Yi, J. Effect of chemical stabilizers in silver nanoparticle suspensions on nanotoxicity. *Bull. Korean Chem. Soc.,* **2011**, *32*(2), 613-619.
[http://dx.doi.org/10.5012/bkcs.2011.32.2.613]

[42]   Bélteky, P.; Rónavári, A.; Igaz, N.; Szerencsés, B.; Tóth, I.Y.; Pfeiffer, I.; Kiricsi, M.; Kónya, Z. Silver nanoparticles: aggregation behavior in biorelevant conditions and its impact on biological activity. *Int. J. Nanomedicine,* **2019**, *14*, 667-687.
[http://dx.doi.org/10.2147/IJN.S185965] [PMID: 30705586]

[43]   Tharmaraj, V.; Yang, J. Sensitive and selective colorimetric detection of $Cu^{(2+)}$ in aqueous medium *via* aggregation of thiomalic acid functionalized Ag nanoparticles. *Analyst (Lond.),* **2014**, *139*(23), 6304-6309.
[http://dx.doi.org/10.1039/C4AN01449A] [PMID: 25316548]

[44]   Porcaro, F.; Carlini, L.; Ugolini, A.; Visaggio, D.; Visca, P.; Fratoddi, I.; Venditti, I.; Meneghini, C.; Simonelli, L.; Marini, C.; Olszewski, W.; Ramanan, N.; Luisetto, I.; Battocchio, C. Synthesis and Structural Characterization of Silver Nanoparticles Stabilized with 3-Mercapto-1-Propansulfonate and 1-Thioglucose Mixed Thiols for Antibacterial Applications. *Mater. (MDPI),* **2016**, *9*, Article ID. 1028.

[45]   Prasher, P.; Sharma, M.; Singh, S.P.; Rawat, D.S. Barbiturate derivatives for managing multifaceted oncogenic pathways: A mini review. *Drug Dev. Res.,* **2020**.
[http://dx.doi.org/10.1002/ddr.21761] [PMID: 33210368]

[46]   Rudakiya, D.M.; Pawar, K. Bactericidal potential of silver nanoparticles synthesized using cell-free extract of Comamonas acidovorans: *in vitro* and *in silico* approaches. *3 Biotech.,* **2017**, *7*, Article 92.

[47]   Veerapandian, M.; Yun, K. Functionalization of biomolecules on nanoparticles: specialized for antibacterial applications. *Appl. Microbiol. Biotechnol.,* **2011**, *90*(5), 1655-1667.
[http://dx.doi.org/10.1007/s00253-011-3291-6] [PMID: 21523475]

[48]   Ansari, M.A.; Asiri, S.M.M.; Alzohairy, M.A.; Alomary, M.N.; Almatroudi, A.; Khan, F.A. Biofabricated fatty acid-capped silver nanoparticles as potential antibacterial, antifungal, antibiofilm, and anticancer agents. *Pharmaceuticals (Basel),* **2021**, *14*(2), 139.
[http://dx.doi.org/10.3390/ph14020139] [PMID: 33572296]

[49]   Ramesh, S.; Grijalva, M.; Debut, A.; de la Torre, B.G.; Albericio, F.; Cumbal, L.H. Peptides conjugated to silver nanoparticles in biomedicine - a "value-added" phenomenon. *Biomater. Sci.,* **2016**, *4*(12), 1713-1725.
[http://dx.doi.org/10.1039/C6BM00688D] [PMID: 27748772]

[50]   Pal, I.; Bhattacharyya, D.; Kar, R.K.; Zarena, D.; Bhunia, A.; Atreya, H.S. A peptide-nanoparticle system with improved efficacy against multidrug resistant bacteria. *Sci. Rep.,* **2019**, *9*(1), 4485.
[http://dx.doi.org/10.1038/s41598-019-41005-7] [PMID: 30872680]

[51]   Halawani, E.M.; Hassan, A.M.; Gad El-Rab, S.M.F. Nanoformulation of biogenic cefotaxime-conjugated-silver nanoparticles for enhanced antibacterial efficacy against multidrug-resistant bacteria and anticancer studies. *Int. J. Nanomedicine,* **2020**, *15*, 1889-1901.
[http://dx.doi.org/10.2147/IJN.S236182] [PMID: 32256066]

[52]   Wang, N.; Zhang, D.; Deng, X.; Sun, Y.; Wang, X.; Ma, P.; Song, D. A novel surface plasmon

resonance biosensor based on the PDA-AgNPs-PDA-Au film sensing platform for horse IgG detection. *Spectrochim. Acta A Mol. Biomol. Spectrosc.,* **2018**, *191*, 290-295.
[http://dx.doi.org/10.1016/j.saa.2017.10.039] [PMID: 29054067]

[53]   Yuce, M.; Kurt, H. How to make nanobiosensors: surface modification and characterization of nanomaterials for biosensing applications. *RSC Advances,* **2017**, *7*(78), 49386-49403.
[http://dx.doi.org/10.1039/C7RA10479K]

[54]   Saleh, T.; Ahmed, E.; Yu, L.; Hussein, K.; Park, K-M.; Lee, Y-S.; Kang, B-J.; Choi, K-Y.; Choi, S.; Kang, K-S.; Woo, H-M. Silver nanoparticles improve structural stability and biocompatibility of decellularized porcine liver. *Artif. Cells Nanomed. Biotechnol.,* **2018**, *46*(sup2), 273-284.
[http://dx.doi.org/10.1080/21691401.2018.1457037] [PMID: 29587547]

[55]   Sonawane, M.D.; Nimse, S.B. Surface modification chemistries of materials used in diagnostic platforms with biomolecules. *J. Chem.,* **2016**, *2016*: 9241378.
[http://dx.doi.org/10.1155/2016/9241378]

[56]   Wickramathilaka, M.P.; Tao, B.Y. Characterization of covalent crosslinking strategies for synthesizing DNA-based bioconjugates. *J. Biol. Eng.,* **2019**, *13*(1), 63.
[http://dx.doi.org/10.1186/s13036-019-0191-2] [PMID: 31333759]

[57]   Thomas, R.K.; Sukumaran, S.; Sudarsanakumar, C. An insight into the comparative binding affinities of chlorogenic acid functionalized gold and silver nanoparticles with ctDNA along with its cytotoxicity analysis. *J. Mol. Liq.,* **2019**, *287*: 110911.
[http://dx.doi.org/10.1016/j.molliq.2019.110911]

[58]   Zhang, X.; Servos, M.R.; Liu, J. Fast pH-assisted functionalization of silver nanoparticles with monothiolated DNA. *Chem. Commun. (Camb.),* **2012**, *48*(81), 10114-10116.
[http://dx.doi.org/10.1039/c2cc35008d] [PMID: 22951627]

[59]   Yu, L.; Li, N. Binding Strength of Nucleobases and Nucleosides on Silver Nanoparticles Probed by a Colorimetric Method. *Langmuir,* **2016**, *32*(22), 5510-5518.
[http://dx.doi.org/10.1021/acs.langmuir.6b01192] [PMID: 27191896]

[60]   AbdulHalim, L.G.; Kothalawala, N.; Sinatra, L.; Dass, A.; Bakr, O.M. Neat and complete: thiolate-ligand exchange on a silver molecular nanoparticle. *J. Am. Chem. Soc.,* **2014**, *136*(45), 15865-15868.
[http://dx.doi.org/10.1021/ja508860b] [PMID: 25345688]

[61]   Jaskolska, D.E.; Brougham, D.F.; Warring, S.L.; McQuillan, J.; Rooney, J.S.; Gordon, K.C.; Meledandri, C.J. Competition-Driven Ligand Exchange for Functionalizing Nanoparticles and Nanoparticle Clusters without Colloidal Destabilization. *ACS Appl. Nano Mater.,* **2019**, *2*(4), 2230-2240.
[http://dx.doi.org/10.1021/acsanm.9b00183]

[62]   López-Lorente, Á.I.; Soriano, M.L.; Valcárcel, M. Analysis of citrate-capped gold and silver nanoparticles by thiol ligand exchange capillary electrophoresis. *Mikrochim. Acta,* **2014**, *181*(15-16), 1789-1796.
[http://dx.doi.org/10.1007/s00604-014-1218-5]

[63]   Sardar, R.; Park, J-W.; Shumaker-Parry, J.S. Polymer-induced synthesis of stable gold and silver nanoparticles and subsequent ligand exchange in water. *Langmuir,* **2007**, *23*(23), 11883-11889.
[http://dx.doi.org/10.1021/la702359g] [PMID: 17918982]

[64]   Ko, Y.; Baek, H.; Kim, Y.; Yoon, M.; Cho, J. Hydrophobic nanoparticle-based nanocomposite films using *in situ* ligand exchange layer-by-layer assembly and their nonvolatile memory applications. *ACS Nano,* **2013**, *7*(1), 143-153.
[http://dx.doi.org/10.1021/nn3034524] [PMID: 23214437]

[65]   Lv, Y.; Lu, M.; Yang, Y.; Yin, Y.; Zhao, J. Electrochemical detection of intracellular glutathione based on ligand exchange assisted release of DNA-templated silver nanoparticles. *Sensor. Sens. Actuators B Chem.,* **2017**, *244*, 151-156.
[http://dx.doi.org/10.1016/j.snb.2016.12.136]

[66] Cho, J.S.; Jang, W.; Kim, S.P.B.K.; Wang, D.H. Thermally stable propanethiol–ligand exchanged Ag nanoparticles for enhanced dispersion in perovskite solar cells *via* an effective incorporation method. *J. Ind. Eng. Chem.,* **2018**, *61*, 71-77.
[http://dx.doi.org/10.1016/j.jiec.2017.12.002]

[67] Lee, Y.J.; Kim, N.R.; Lee, C.; Lee, H.M. Uniform thin film electrode made of low-temperatur--sinterable silver nanoparticles: optimized extent of ligand exchange from oleylamine to acrylic acid. *J. Nanopart. Res.,* **2017**, *19*(2), 32.
[http://dx.doi.org/10.1007/s11051-016-3720-7]

[68] Merg, A.D.; Zhou, Y.; Smith, A.M.; Millstone, J.E.; Rosi, N.L. Ligand exchange for controlling the surface chemistry and properties of nanoparticle superstructures. *ChemNanoMat,* **2017**, *3*(10), 745-749.
[http://dx.doi.org/10.1002/cnma.201700171]

[69] Long, Y-M.; Hu, L-G.; Yan, X-T.; Zhao, X-C.; Zhou, Q-F.; Cai, Y.; Jiang, G-B. Surface ligand controls silver ion release of nanosilver and its antibacterial activity against *Escherichia coli. Int. J. Nanomedicine,* **2017**, *12*, 3193-3206.
[http://dx.doi.org/10.2147/IJN.S132327] [PMID: 28458540]

[70] Suzuki, Y. Development of fluorescent reagent based on ligand exchange reaction for the highly sensitive and selective detection of dopamine in the serum. *Sensors (Basel),* **2019**, *19*(18), 3928.
[http://dx.doi.org/10.3390/s19183928] [PMID: 31547244]

[71] Song, J.; Zhu, Y.; Zhang, J.; Yang, J.; Du, Y.; Zheng, W.; Wen, C.; Zhang, Y.; Zhang, L. Encapsulation of AgNPs within Zwitterionic Hydrogels for Highly Efficient and Antifouling Catalysis in Biological Environments. *Langmuir,* **2019**, *35*(5), 1563-1570.
[http://dx.doi.org/10.1021/acs.langmuir.8b02918] [PMID: 30563342]

[72] Qasim, M.; Udomluck, N.; Chang, J.; Park, H.; Kim, K. Antimicrobial activity of silver nanoparticles encapsulated in poly-*N*-isopropylacrylamide-based polymeric nanoparticles. *Int. J. Nanomedicine,* **2018**, *13*, 235-249.
[http://dx.doi.org/10.2147/IJN.S153485] [PMID: 29379284]

[73] Zaheer, Z.; Aazam, E.S.; Hussain, S. Reversible encapsulation of silver nanoparticles into the helix of amylose (water-soluble starch). *RSC Advances,* **2016**, *6*(65), 60513-60521.
[http://dx.doi.org/10.1039/C6RA09319A]

[74] Khan, Z.; Al-Thabaiti, S.A. Biogenic silver nanoparticles: Green synthesis, encapsulation, thermal stability and antimicrobial activities. *J. Mol. Liq.,* **2019**, *289*: 111102.
[http://dx.doi.org/10.1016/j.molliq.2019.111102]

[75] Sathiyaseelan, A.; Saravanakumar, K.; Mariadoss, A.V.A.; Wang, M-H. Biocompatible fungal chitosan encapsulated phytogenic silver nanoparticles enhanced antidiabetic, antioxidant and antibacterial activity. *Int. J. Biol. Macromol.,* **2020**, *153*, 63-71.
[http://dx.doi.org/10.1016/j.ijbiomac.2020.02.291] [PMID: 32112842]

[76] Liu, X.; Hao, W.; Lok, C-N.; Wang, Y.C.; Zhang, R.; Wong, K.K.Y. Dendrimer encapsulation enhances anti-inflammatory efficacy of silver nanoparticles. *J. Pediatr. Surg.,* **2014**, *49*(12), 1846-1851.
[http://dx.doi.org/10.1016/j.jpedsurg.2014.09.033] [PMID: 25487498]

[77] Silva, H.F.O.; Lima, K.M.G.; Cardoso, M.B.; Oliveira, J.F.A.; Melo, M.C.N.; Anna, C.S.; Eugenio, M.; Gasparotto, L.H.S. Doxycycline conjugated with polyvinylpyrrolidone-encapsulated silver nanoparticles: a polymer's malevolent touch against *Escherichia coli. RSC Advances,* **2015**, *5*(82), 66886-66893.
[http://dx.doi.org/10.1039/C5RA10880B]

[78] Ghodake, G.; Shinde, S.; Saratale, R.G.; Kadam, A.; Saratale, G.D.; Patel, R.; Kumar, A.; Kumar, S.; Kim, D-Y. Whey peptide-encapsulated silver nanoparticles as a colorimetric and spectrophotometric probe for palladium(II). *Microchim. Acta,* **2019**, *186*, Article ID. 763.

[79]   Taheri, S.; Cavallaro, A.; Christo, S.N.; Majewski, P.; Barton, M.; Hayball, J.D.; Vasilev, K. Antibacterial Plasma Polymer Films Conjugated with Phospholipid Encapsulated Silver Nanoparticles. *ACS Biomater. Sci. Eng.,* **2015**, *1*(12), 1278-1286.
[http://dx.doi.org/10.1021/acsbiomaterials.5b00338] [PMID: 33429675]

[80]   Fujiwara, K.; Kuwahara, Y.; Sumida, Y.; Yamashita, H. Synthesis of Ag nanoparticles encapsulated in hollow silica spheres for efficient and selective removal of low-concentrated sulfur compounds. *J. Mater. Chem. A Mater. Energy Sustain.,* **2017**, *5*(48), 25431-25437.
[http://dx.doi.org/10.1039/C7TA08918J]

[81]   Pavoski, G.; Baldisserotto, D.L.S.; Maraschin, T.; Brum, L.F.W.; Santos, C.; Santos, J.H.Z.; Brandelli, A.; Galland, G.B. Silver nanoparticles encapsulated in silica: Synthesis, characterization and application as antibacterial fillers in the ethylene polymerization. *Eur. Polym. J.,* **2019**, *117*, 38-54.
[http://dx.doi.org/10.1016/j.eurpolymj.2019.04.055]

[82]   Karunamuni, R.; Naha, P.C.; Lau, K.C.; Al-Zaki, A.; Popov, A.V.; Cormode, D.P.; Delikatny, E.J.; Tsourkas, A.; Maidment, A.D.A. *Development of silica-encapsulated silver nanoparticle as contrast agents intended for dual-energy mammography.,* **2016**.
[http://dx.doi.org/10.1007/s00330-015-4152-y]

CHAPTER 5

Applications of Bioconjugated Silver Nanoparticles

Abstract: The bioconjugated AgNPs present various applications starting from advanced molecular medicine agents, drug delivery vectors, gene transfection, bioimaging and contrast enhancement agents, and as anticancer chemotherapeutics. Mainly, the theranostic applications of AgNPs combined with the therapeutic effect of the conjugated biomolecule result in advanced applications. The most recent state-o--the-art applications of bioconjugated or biogenic AgNPs include the magnetic imaging of vital life processes, cellular internalization of bioactive ingredients or therapeutics, DNA repair and gene transfection in molecular medicine for correcting the anomalies at the genetic level. The conjugation with AgNPs causes a considerable improvement in the therapeutic efficacy of anticancer chemotherapeutics owing to targeted release coupled with the physiological changes associated with the interaction of biological systems with AgNPs that triggers redox stress and heightened cellular response. The present chapter deals with the applications of AgNPs conjugated to biological ingredients or biogenic AgNPs in various fields.

Keywords: Anticancer effect, Bioimaging, Drug delivery, Gene transfection, Molecular medicine.

1. APPLICATION IN MOLECULAR MEDICINE

The AgNPs owing to their unique features, present pervasive biological applications. Mainly, the utilization of AgNPs in molecular medicine as drug delivery vehicles, diagnostic probes, contrast agents, and as gene transfection agents offer the most advanced applications [1, 2]. The ability of AgNPs to pass through the membrane barriers promote their drug delivery applications, the optoelectronic properties enable their applications as diagnostic probes, magnetic and surface plasmonic properties offer utility as contrast agents, whereas the ability to bioconjugate to biomolecules such as nucleic acids, antibodies allow the gene transfection properties of AgNPs [3]. The fine size of AgNPs enables their accumulation at the target site. This property of AgNPs presents applications for evading the drug efflux by the vulnerable drug molecules [4 - 6]. The pharmaceuticals conjugated to AgNPs hence maintaining an optimal concentration at the target site that exceeds the efflux effect created by the multi drug resistant cells. The conjugation of AgNPs to biomolecules by the strategies such as EDC-coupling, sulfo-NHS coupling, and click chemistry enable the formation of strong bonds that provide enhanced stability to metal-based nanosystems directed at biological applications [7, 8]. In addition, these strategies discourage the direct exposure of AgNPs and Ag^+ ions to the biological systems that may prove hazardous by instigating long-term deleterious effects. The AgNPs improve the susceptibility of oncogenic cells towards radiosensitization for enhanced performance in radiotherapy [9 - 11]. The conjugation of antibodies to AgNPs serves the same purpose of improving the vulnerability of cancer cells towards radiotherapy [12]. Liu *et al.* (2018) reported an enhanced efficacy of radiotherapy in hypoxic glioma cells in the presence of AgNPs. The AgNPs displayed IC50 = 27.53 µg/mL, and 30.32 µg/mL towards hypoxic C6 cells and U251 cells, respectively. The sensitization enhancement ratio depicted the higher efficacy of AgNPs in ameliorating the radiosensitization in hypoxic cells compared to the normoxic cells, mainly by promoting apoptosis and enhanced destructive autophagy. The cancer cell-lines treated with AgNPs and X-rays displayed higher levels of the reactive oxygen species, compared to the cells treated with X-rays only. The elevated ROS levels cause an increase in autophagy that further leads to the destruction of cytoplasm and vital organelles resulting in the loss of cellular functions. The AgNPs induced autophagy results in the development of a primary response towards tress stimuli that instigates apoptosis *via* regulatory proteins, such as caspase, Bcl-2, P53, and ATG, or ERK (extracellular signal-regulated kinase) and JNK (c-Jun N-terminal kinase) pathways [13]. Yu *et al.* (2017) reported the sensitization of nasopharyngeal carcinoma cells toward radiotherapy in the presence of AgNPs coupled to anti-EGFR antibodies. The AgNPs inhibited the proliferation of carcinoma cells in a time- and dose-dependent manner. The flow cytometry analysis indicated the induction of apoptosis in carcinoma cells by arresting the cell cycle in G2 phase, which improved on conjugating the AgNPs with anti-EGFR antibodies. The clonogenic assays suggested enhanced sensitization of carcinoma cells towards irradiation on treatment with antibody-conjugated AgNPs, with sensitizer enhancement ratios of 1.405±0.033 and 1.610±0.012 respectively for antibody-conjugated AgNPs and unconjugated AgNPs. Similarly, the western blotting analysis confirmed that the combination of AgNPs/ antibody-conjugated AgNPs with X-ray irradiation downregulated the expression of DNA damage or DNAA repair proteins including Rad-51, Ku-70 and Ku-80; with antibody-conjugated AgNPs displaying a more pronounced effect [14]. Zhao *et al.* (2019) reported an enhancement in the radiosensitization of glioma cells on treatment with AgNPs-conjugated with polyethylene glycol and Aptamer As1411. The spherical shaped nanoparticles with a diameter of 18 nm effectively targeted C6 glioma cells, while sparing the normal human microvascular endothelial cells that result in selective attenuation of cancer cells without affecting nearby healthy cells. The nanoparticles effectively internalized tumor cells and penetrated the core of tumor spheroids for an optimal effect. The conjugation of Aptamer As1411 with AgNPs produced a superior radiosensitization effect, compared to their PEGylation, with the former displaying a higher rate of apoptosis and cell death. The aptamer-conjugated AgNPs are mostly localized in the C6 glioma cell cytoplasm, while they are distributed outside the cell or in the plasm membrane in the non-malignant HMEC-1 cells. Mainly, the amount of cell surface nucleolin determines this differential AgNPs uptake by the cells. The nucleolin mainly expresses on the surface of tumor cells and not in healthy cells. The arginine-glycine-glycine domain of nucleolin selectively binds to the G-quadruplex of As1411, thereby initiating the internalization of aptamer conjugated-AgNPs [9]. Zhao *et al.* (2021) reported enhanced radiosensitivity of glioma caused by increased accumulation of aptamer As1411 and verapamil conjugated AgNPs in cancer cells. The mixing of aptamer-conjugated AgNPs and verapamil conjugated BSA coated AgNPs in ratio 19:1 resulted in a significant upsurge in the intracellular nanoparticles *via* As1411-mediated active targeting and inhibition of P-glycoproteins. The *in vivo* and *in vitro* analysis showed that the aptamer/ verapamil bioconjugated AgNPs formulations mixed at 19:1 significantly inhibited TrxR activity and demonstrated a marked efficiency compared to the individual aptamer-conjugated or verapamil-conjugated AgNPs [15]. Jyoti *et al.* (2020) reported radiosensitizing and cytotoxic potential of AgNPs against HepG-2 cells. The spherical shaped AgNPs with average diameter 5-40 nm synthesized from *Picrasma quassioides* displayed a negative zeta potential with marked anticancer properties against U251 glioblastoma cells. The cancer cell administered with AgNPs displayed a significant inhibition at a gamma radiation dose of 6 Gγ, and increased the autophagy for glioma treatment. Mainly the internalization of AgNPs in cancer cells leads to their annihilation. The AgNPs pose hazardous effects on the nitrogen bases and phosphate groups of cancer cell DNA by physicochemical interactions that cause the death of cancer cells [16]. Fathy *et al.* (2020) reported the radiosensitizing properties of biosynthesized AgNPs using thymoquinone as reducing and capping agent. Reportedly, the AgNPs displayed radiosensitizing effect against highly aggressive MDA-MB-231 mammary adenocarcinoma, where the nanoparticles caused DNA damage, thereby leading to the enhanced killing of cancer cells. Notably, as the size of AgNPs increases, the secondary ionizing events caused by the interactions of AgNPs with radiation causes a significant reduction in the accumulated dose around the nanoparticles. Similarly, the concentration of nanoparticles played a significant effect on the radio-sensitizing properties due to enhanced internalization of nanoparticles that lowered cancer cell viability due to increased possibility of photoelectric interactions in the cancer cells [17]. Wu *et al.* (2016) reported autophagy induced by AgNPs by radiation enhancement caused by the generation of the reactive oxygen species by AgNPs.

The ROS generated by nanoparticles played a critical role in the induction of autophagy and radiosensitizing effect in the U251 cells causing their death and apoptosis. On the activation of autophagy, the lipidation of the cytoplasmic forms of LC3 and its recruitment to autophagosomes membrane causes cell apoptosis [18]. Swanner *et al.* (2015) displayed the cytotoxicity and radiosensitizing effects of AgNPs towards Triple-negative breast cancer cells (TNBC). The AgNPs capped with PVP induced DNA damage by inciting oxidative stress, while displaying lower cytotoxicity to healthy cells. The intratumoral injection of nanoparticles decreased the growth of TNBC xenograft and improved the effect of radiation therapy *in vivo* [19]. Liu *et al.* (2013) reported the radiation sensitization properties of AgNPs for the treatment of glioma. The intratumoral injection of AgNPs combined with a single dose of ionizing radiation at 10 Gγ proved highly efficacious in the annihilation of glioma. The treatment of gliomas with AgNPs and radiation resulted in antiproliferative and proapoptotic effect that caused tumor cell death after radiotherapy. The combination of radiotherapy and AgNPs caused a considerable improvement in the survival time without apparent toxicity [20].

1.1. Application in Drug Delivery

The AgNPs play an important role in drug delivery applications mainly due to their ability to internalize the cells and conjugate with a range of chemical molecules. The AgNPs based drug delivery systems effectively overcome the intricacies associated with contemporary anticancer and antimicrobial drugs, including multidrug resistance and antimicrobial resistance. Qiu *et al.* (2018) reported the drug delivery application of hybrid formulations of polymers with AgNPs. The conjugation of redox responsive copolymer P[(2-((2-((camptot-hecin)-oxy)ethyl)disulfanyl)ethylmethacrylate)-co-(2-(d-galactose)methylmeth-acryl-ate)] (P(MACPTS-co-MAGP)) with AgNPs effectively assisted in the delivery of the anticancer drug camptothecin, where the fluorescence of the drug molecule enabled the monitoring of drug release. The conjugation of the drug molecule occurs with side chains of the polymer *via* redox responsive disulfide linkages. The presence of glutathione causes the cleavage of disulfide linkages to set-free the conjugated drug molecules. Notably, these hybrid nanoparticles suppress the proliferation of HeLa cells, which provides added advantage in drug delivery applications [21]. Nezami *et al.* (2020) reported pH-sensitive free AgNPs composite as drug delivery systems for methyl prednisolone. The AgNPs nanocomposite presented a superior drug release, which improved the amount of AgNPs in the polymer matrix until 3.3%. Further enhancement in the AgNPs content caused a lowering of the drug release profile by AgNPs composites. Similarly, the AgNPs nanocomposites achieve a maximum drug release at pH 7.4

for 9 hours, while in the acidic medium at pH 1.6, the drug release by the AgNPs composite lowered mainly due to the shrinkage of the polymer matrix in acidic media. The drug-release kinetics indicated a diffusion-swelling controlled drug-release process adopted by the AgNPs composites that present applications in oral delivery of drug molecules [22]. Gul *et al.* (2020) reported drug evaluation of biogenic AgNPs for anticancer *Euphorbia dracunculoides* Lam plant extract. The biogenic AgNPs demonstrated DPPH-free radical scavenging and antioxidant properties. The prolonged ingestion of the maximum dose of the drug loaded drug delivery system at 300 mg/ Kg displayed non-cytotoxic properties and caused a gradual rise in the levels of TNF-α, which caused a reduction in the inflammatory lymphocytes and monocytes during the anticancer treatment. Importantly, the reported biogenic AgNPs with a diameter in the range 36.66 nm displayed marked stability in various solvents, including the deionized water and phosphate-buffered saline solution at 4°C for over one month [23]. Benyettou *et al.* (2015) reported dual delivery application of AgNPs for doxorubicin, and alendronate to the cancer cells. The coating of AgNPs with bisphosphonate alendronate with free primary ammonium group served as the main site for tethering anticancer drug doxorubicin. The functionalization of the primary ammonium group of alendronate with rhodamine B *via* amide bond, attachment of doxorubicin through imine bond proved highly advantageous for pH-triggered controlled release of the drugs *in vivo*. The imine bond conjugation of AgNPs with doxorubicin enabled acid-mediated intracellular release of the drug with a significant anticancer potency compared to individual doxorubicin. Mainly, the cellular uptake of the drug loaded AgNPs occurred *via* clatharin-mediated endocytosis or macropinocytosis, where the pH-responsive release of the drug molecules in late lysosomes or endosomes cause their even distribution in the cytosol [24]. Qiu *et al.* (2017) reported drug delivery applications of AgNPs loaded with pH-sensitive camptothecin-loaded polymeric prodrugs. The drug links to the polymeric side chains by acid-sensitive β-thiopropionate bond prepared by RAFT polymerization. The *in vitro* investigations suggested that the fluorescence of drug molecules at pH 7.4 in buffer solution quenches owing to the nanoparticle surface-energy transfer, however, in the acidic medium the fluorescence recovery takes place followed by a gradual release of drug molecules *via* degradation of acid-sensitive bonds. The internalization of drug molecules loaded nanoparticle in lysosomes causes blue fluorescence due to drug release in acidic media, which increases with incubation time. Importantly, the cytotoxicity of drug carrying nanoparticle depended on the activity of acid-sensitive bonds [25]. Bhanumathi *et al.* (2018) reported the drug carrying capacity and anticancer potency of folic acid- and Berberine-loaded AgNPs that successfully regulated AKT-ERK pathway in breast cancer. The encapsulation of berberine takes place on citrate-capped AgNPs *via* electrostatic interactions, which further conjugated to polyethylene glycol functionalized folic acid *via* hydrogen bonding interactions. The drug loaded nanosystem aggregates at the tumor site for releasing the cargo drug molecules, thereby achieving an optimal therapeutic effect. In addition, the nanocarriers promoted the induction of apoptosis by the generation of reactive oxygen species and condensed nuclei in the target cancer cells. The western blotting analysis indicated the modulation in the expression of AKT, Ras, Raf, VEGF, Bcl-2, HIF-1α, PI3K, caspase-3 and 9, cytochrome c, and Bax pathways that contributed towards the enhancement in apoptosis, resulting in a marked restriction of tumor progression [26]. Fagbenro *et al.* (2019) reported drug delivery applications of egg protein stabilized AgNPs for hesperidin. The hydrophilic nanosystem displayed high loading efficiency for the hesperidin due to the presence of anionic groups, and enhanced their antibacterial potency against the multidrug resistant strains of *S. aureus*. The electrostatic complexation of silver ions with oppositely charged free thiol groups present in the proteins assisted in the nucleation processes of metallic silver with sodium ascorbate leading to the formation of spherical AgNPs caused by isotropic growth due to the presence of –OH, and –NH groups. The presence of thiol. Hydroxyl, carboxyl, and amine groups in the proteins donate an electron to the ionic silver for generating AgNPs by acting as reducing agents. The nanosystem notably showed high stability in blood plasma where the controlled release of the loaded drug molecules takes place [27]. Rozalen *et al.* (2020) reported the size-controlled AgNPs in the delivery of methotrexate and for enhanced anticancer activity towards colon and lung cancer-cell lines. The loading of the drug molecule occurred on citrate caped AgNPs *via* the chemisorption of –COOH groups of the former. The nanosystem achieved a controlled and sustained release of the loaded drug molecules, compared to the free methotrexate that resulted in the improved pharmacokinetics of the cargo drug molecules. Reportedly, the free drug molecules release in 3 hours while following the first order kinetics, whereas their loading on AgNPs caused a delayed release by 5 hour. The AgNPs caused a faster drug release in the beginning, while after 24 hours, a drug release plateau reached whereby the rate of release became constant [28]. Li *et al.* (2016) reported the drug delivery applications of polyethylenimine-capped AgNPs for the delivery of paclitaxel to trigger apoptosis in the HepG2 cells. The surface capping of AgNPs with polyethylenimine and paclitaxel attacked the target cancer cells with high specificity by inducing apoptosis, enhancing cytotoxicity, depletion of the potential of mitochondrial membrane, activation of caspase-3, DNA fragmentation, translocation of phosphatidylserine, and the cleavage of poly(ADP-ribose) polymerase. Importantly, the drug delivery nanosystem enhanced the activation of signaling pathways including MAK, p53, and AKT followed by the generation of oxidative stress that played a key role in affecting the cancer cell viability [29]. Kalindemirtas *et al.* (2021) reported selective cytotoxicity of paclitaxel bonded AgNPs on various cell lines, including MCF-7, MDA-MB-231, 4T1, Saos-2, and the non-cancerous HUVEC cells. The conjugation of paclitaxel on AgNPs proved ten-times more effective compared to the paclitaxel alone for targeting the cancer cells. Interestingly, the nanoformulations proved non-toxic towards the HUVEC cell that further validated their selectivity towards the cancer cells [30]. Muhammad *et al.* (2020) presented the functionalization of paclitaxel on polydopamine capped AgNPs to provide an enhanced anticancer effect on the human cancer cell lines. The nanosystem improved the cellular uptake of the anticancer drug *via* NR1-receptor interactions, promoted the pH-sensitive drug release, and demonstrated synergistic anticancer effect with AgNPs. The nanocarriers exhibited the induction of apoptosis, thereby causing lysis of cancer cell membrane, irreversible damage to the nucleus, mitochondrial dysfunction, and cleavage of host cell DNA. In addition, the nanosystem effectively regulated the mitochondria-based apoptosis by the activation of pro-apoptotic P53, and caspase 3 [31]. Mohamed *et al.* (2020) reported hybrid chitosan AgNPs loaded with doxorubicin for offering anticancer properties. The nanosystem demonstrated pH-responsive sustained drug release on human breast cancer cells and subcutaneous tumors by displaying an enhanced toxicity towards the cancer cells [32].

1.2. Application in Bioimaging

The plasmonic, electrical and magnetic properties of AgNPs support their theranostic applications. The optical and electrical properties favor the bioimaging applications, while the magnetic properties of AgNPs offer applications as image contrast enhancers or contrast agents in non-invasive techniques such as magnetic resonance imaging, and positron emission tomography [33]. The AgNPs improve the detection limit of the target due to their thermochemical, and catalytic stability. The AgNPs attach to the surface of the sensing surface mainly *via* electrostatic interactions; however, suitable chemical modifications further improve the attachment of AgNPs on the sensing surface. The AgNPs possess a lower local refractive index compared to the molecules, which further improves on conjugation of AgNPs with biomolecules. Similarly, the presence of a suitable surface coating improves the detection of biomolecules. The luminescent AgNPs present applications in the imaging of leukemia and neural stem cells while penetrating to the cell membrane of the target cells. Kravets *et al.* (2016) performed the imaging of basophilic leukemia, and neural stem cells by luminescent AgNPs functionalized with glycine dimers. The spectral peaks for intrinsic and extrinsic emission of AgNPs undergo red shift as the size of AgNPs increased. Similarly, the intrinsic photoluminescence of AgNPs weakened and separated spectrally from the photoluminescence of glycine dimer ligands. The quantum yield of the AgNPs-ligand system depended on the size of nanoparticles, whereas the spectral position of ligand emission was independent of the particle size, mainly due to the augmentation of ligand emission by the strength of the local electrical field. The AgNPs effectively penetrated the cell membranes of the target cells that enabled the bioimaging of the latter [34]. Nie *et al.* (2018) reported the bioimaging applications of polyvinyl pyrrolidone-coated fluorescent AgNPs. The surface morphology and fluorescence property of AgNPs determined their imaging applications. Mainly, the remarkable chemical stability of AgNPs, considerable biocompatibility, and bright photoluminescence properties enabled the labelling of cancer cells and in the imaging of biological environment [35]. Mondal *et al.* (2018) reported the imaging applications of AgNPs as contrast agents using swept-source optical coherence tomography. The AgNPs serve as an exogenous contrast agent for *in vitro* imaging of the cells. The application of nanoparticles resulted in contrast enhancement and for the enhancement in the imaging depth. The OCT images were taken in the presence of AgNPs efficiently localized in the underlying microstructures in the target cells that provided images of the latter [36]. Kim *et al.* (2011) evaluated the toxicity of AgNPs by performing metabolomics based on high-resolution magic angle spinning nuclear magnetic resonance. Reportedly, the AgNPs influenced the concentration of liver cell metabolites. The AgNPs augmented the concentration of taurine, lactate, glycine, and glutathione. The treatment with *N*-acetylcysteine however resulted in the recovery of these metabolites [37]. Table **1** illustrates the bioconjugated AgNPs, their morphology and imaging applications.

Table 1. Bioconjugated AgNPs for imaging applications.

Bioconjugated AgNPs	Morphology/Characteristics	Imaging Property	Refs.
Epigallocatechin gallate functionalized AgNPs	Spherical shape, average size 5 nm	Multicolor bioimaging/ luminescent property	[38]
Polyvinyl pyrrolidone conjugated AgNPs	Average diameter 8.7-50 nm	Bright photoluminescence property, labelling of cancer cells	[35]
Multifunctional AgNPs nanohybrids	Average diameter 80 nm	Aggregation induced emission-enhanced photodynamic therapy	[39]
Colloidal solution of AgNPs	Spherical shape, average size 80-100 nm	Contrast agent for imaging of animal tissues, application in swept-source optical coherence tomography	[36]
AgNPs-embedded microbubble	Spherical shape, average size 50 nm	Dual-mode ultrasound and optical imaging probes	[40]

(Table 1) cont.....

Bioconjugated AgNPs	Morphology/Characteristics	Imaging Property	Refs.
Fluorescent Ag nanoconjugates	Spherical shape	Bioimaging of the Cu(ii) scavenging process in human lymphocytes by atomic force microscopy	[41]
AgNPs embedded hybrid organometallic complexes	Spherical shape, average size 6.5 nm	Photo-induced energy transfer, plasmonic effect, optical thermometry	[42]

1.3. Application in Gene Transfection

AgNPs serve as a noble alternative to the non-viral and viral vehicles for the transporting of exogenous DNA to the target cells, mainly due to the ability to conjugate with lipids, peptides, and ligands for crossing the endosomal, and membrane barriers. The fine size of AgNPs and a large surface area for adhesion provide easy transit across the cell membranes, cellular internalization, and integration with intracellular pathways. The higher stability of AgNP with anchored ligands prevents their digestion by the enzymes, which enables the delivery of attached therapeutic or nucleic acid at the deliberated site. Table **2** presents the gene transfection applications of bioconjugated AgNPs.

Table 2. Bioconjugated AgNPs for gene transfection.

Bioconjugated AgNPs	Morphology/Characteristics	Gene Transfection Property	Refs.
H5 DNA plasmid encapsulated in AgNPs	Average diameter 25 nm with a positive charge of +40 ± 6.2 mV	Gene expression profiling in primary duodenal chick cells	[43]
Plasmid DNA conjugated AgNPs	Well dispersed AgNPs, average diameter 10 nm	Efficient delivery of condensed pDNA	[44]
Pulsed plasma surface functionalized AgNPs	Spherical AgNPs, average diameter 15-25 nm	Efficient delivery of pDNA	[45]
Polyethylenimine capped AgNPs	Monodispersed, spherical nanoparticles with positive zeta potential	Delivery of VP1 siRNA to inhibit Enterovirus 71	[46]
Cell-penetrating TAT peptide conjugated AgNPs	Spherical nanoconstructs with positive zeta potential	Gene delivery to epidermal stem cells	[47]
siRNA conjugated to quercetin capped AgNPs	Nanoconstructs with average size 40 nm	Gene silencing in drug resistant *Bacillus subtilis*	[12]
RGDS peptide conjugated AgNPs	Biofunctionalized stable AgNPs with narrow size distribution	Gene transfection properties	[48]
NAD(P)H nitroreductase YfkO reduced AgNPs	Ultrafine AgNPs with average diameter 50 nm	Horizontal gene transfer in Marinomonas strain isolated from Antarctic psychrophilic ciliate Euplotes focardii	[49]

(Table 2) cont.....

Bioconjugated AgNPs	Morphology/Characteristics	Gene Transfection Property	Refs.
AgNPs and silver ions	Spherical AgNPs, average diameter 15 nm	Horizontal transfer of plasmid mediated antibiotic resistant genes by inducing ROS overproduction, and increasing cell membrane permeability	[50]
Ag@SiO$_2$ hollow nanoparticles	Hollow nanostructures with average diameter 500 nm	Nasal vaccination of animals with Newcastle disease virus DNA vaccine, induced high titers of serum antibody, promoted lymphocyte proliferation, increased the expression of IL2, and IFN-γ	[51]
AgNP-pIREGFP-H5 conjugates	Average diameter 25 nm with positive zeta potential	Gene transfection of avian influenza virus H5 DNA plasmid in primary duodenal chick cells	[43]

1.4. Anticancer Applications

The bioconjugated AgNPs present anticancer properties on their anchoring with bioactive compounds or demonstrate synergistic anticancer potential. Pimentel *et al.* (2016) reported soybean agglutinin-conjugated AgNPs for the treatment of breast cancer cells. The breast cancer cells display alterations in the sugar expression patterns on their surface related to the cancer progression and metastasis. The soybean agglutinin effectively identifies these changes, which makes it a desirable system for conjugation to AgNPs for cancer cell targeting. *In vitro* cytotoxicity analysis indicated the inhibition of breast cancer cell-lines MDA-MB-231 and MCF7. Importantly, the nanoencapsulation lowered the cytotoxicity of AgNPs towards the healthy cells that further validated the efficacy of the reported nanosystem in biological applications [52]. Zhao *et al.* (2021) reported the conjugation of AgNPs with aptamer AS1411 and verapamil for enhanced radiosensitivity of tumor cells in glioma. This ensured the inhibition of tumor recurrence and metastasis that occurs due to radioresistance, and promoted increased accumulation of the conjugated aptamer and the drug in tumor cells. The reported nanoconjugate displayed a high efficacy as nano-radiosensitizers in glioma radiotherapy [15]. Karuppaiah *et al.* (2020) reported the bioconjugation of AgNPs with gemcitabine for inhibiting the proliferation in human triple-negative metastatic breast cancer cells lines MDA-MB-453. Gemcitabine adsorbs on AgNPs surface by electrostatic interactions for form conjugates, which dissociates in the aqueous medium to release the free drug molecules. The reported conjugates significantly enhanced the cytotoxicity of the cargo drug towards the target cancer cells. The reported nanoconjugates displayed lower lymphatic drainage and enhanced accumulation in the tumor sites, which results in increased toxicity towards the tumor cells [53]. Patra *et al.* (2018) reported the anticancer properties of bioconjugated AgNPs prepared using the leaf extract of *Saraca asoca*. The phytochemicals present in the plant extract served as both reducing and capping/ stabilizing agent. Mainly, the functional groups such as ester, alcoholic and phenolic groups caused the synthesis of AgNPs. The prepared bioconjugated AgNPs demonstrated considerable anticancer activity against prostate cancer cell lines with significant biocompatibility [54]. Vijayan *et al.* (2020) presented the anticancer investigations on the chitosan conjugates of biogenic AgNPs on human cervical carcinoma (Si Ha) and human adenocarcinoma (MDA MB) cell lines. The chitosan-bioconjugated AgNPs induced antiproliferative and cell apoptosis-inducing activities, and caused an upregulation of p53 and p38 genes that offered anticancer properties. Furthermore, the DNA fragmentation studies, caspase 7 and 9 assays, and Ethidium bromide/ acridine orange staining experiments explained the tumor cell-death mechanism by chitosan conjugated AgNPs [55]. Majeed *et al.* (2019) reported anticancer activity of bioengineered AgNPs capped with bovine serum albumin in the inhibition of breast, bone and intestine colon-cancer cell lines. The conjugated AgNPs displayed marked anticancer activity against MCF-7, HCT-116, and MG-63 cancer cell lines, with minimal cytotoxicity towards the normal 3T3 skin fibroblast cells. Moreover, the conjugated AgNPs displayed an enhanced release of LDH, which confirmed the augmentation in cancer cell apoptosis. Further confirmation by DNA analysis indicated a complete fragmentation of AgNPs treated cell, hence validating the anticancer properties of the latter [56]. Table **3** presents the anticancer applications of bioconjugated AgNPs.

Table 3. Bioconjugated AgNPs and their anticancer properties.

Bioconjugated AgNPs	Morphology/ Characteristics	Anticancer Property	Ref.
Chitosan-alginate AgNPs	Microporous composites with pore size 50-500 μm	Induce apoptosis in cancer cells lines MDAMB-231	[57]
Interleukin-10 conjugated AgNPs	Polyvinylpyrrolidone-coated spherical shape, diameter 50 nm	Reduce lipopolysaccharide-induced inflammatory response, cytotoxicity towards cancer cells	[58]
Carboxymethylcellulose-doxorubicin-conjugated AgNPs	Spherical nanocolloids with uniform size distribution, average diameter 10nm	Improve pharmacokinetics of anticancer drug, killing of melanoma cells,	[59]

(Table 3) cont.....

Bioconjugated AgNPs	Morphology/ Characteristics	Anticancer Property	Ref.
Arg-Gly-Asp-Ser (RGDS) peptide conjugated AgNPs	Bimodal distribution of particle size, average diameter 68 nm	Gene transfection in HeLa and A549 cells, Binding with cancer cell DNA, minimal cytotoxicity to healthy cells	[48]
Folic acid-conjugated gemcitabine anchored AgNPs	Nanocomposite	Apoptotic cell death, cytotoxicity towards MDA-MB-453 breast cancer cell lines	[60]
Aptamer As1411 conjugated AgNPs	Spherical, well dispersed nanoparticles with average diameter 18.82 nm	Enhancement in radiosensitization for Glioma irradiation therapy	[9]
Poly-glutamic acid conjugated AgNPs	Biocompatible nanocomposite consisting of Multiwalled carbon nanotubes	Electrochemical sandwich biosensor for MCF-7 human breast cancer cells	[61]
Diterpene and saponins conjugated AgNPs	Spherical shape, Average diameter 156 nm	Cytotoxicity towards human embryonic kidney cells	[62]
Peptide conjugated AgNPs	Spherical, average diameter 35 nm	Target nucleus and cytoplasm of cancer cells, break DNA double strand, arrest G2 phase of cell cycle	[63]
Peptide conjugated AgNPs	Spherical, average diameter 1-100 nm	Target nucleus and cytoplasm of cancer cell lines HSC-3	[64]
Nucleotide (Adenosine triphosphate) functionalized AgNPs	Spherical colloidal particles with an average hydrodynamic diameter of about 15–20 nm	Enzyme responsive anticancer properties against human liver carcinoma cells	[65]
Doxorubicin conjugated AgNPs	Spherical nanoparticles with average diameter 48 nm, positive zeta potential	Sustained release of doxorubicin for an improved anticancer effect	[32]
AgNPs conjugated with cell penetrating peptides	Spherical nanoparticles with average diameter 31.61 nm	Enhanced killing of MCF-7 tumor cells due to an improved internalization to the cells	[66]
Biosynthesized colloidal stabilized AgNPs	Ultrafine nanoparticles with average diameter 2 nm	Induction of apoptosis in HePG-2 cells *via* ROS-mediated signaling pathways including MAP-kinase, and AKT-signaling	[67]

(Table 3) cont.....

Bioconjugated AgNPs	Morphology/ Characteristics	Anticancer Property	Ref.
Biosynthesized colloidal stabilized AgNPs	Spherical nanoparticles with average diameter 35 nm	Increased expression of caspase 3/ 9, BH_3-interacting domain death agonist, and Bax. Downregulation of anti-apoptotic proteins Bcl-2 and Bcl-xl.	[68]
Paclitaxel conjugated AgNPs	Spherical nanoparticles with average diameter 40 nm	Anticancer effect on MDA-MB-231, MCF-7, 4T1, Saos-2, and HUVEC cell lines. Effective in the chemotherapy of osteosarcoma and breast cancer	[30]
Biosynthesized colloidal stabilized AgNPs	Triangular silver nanoplates	Anticancer effect against MCF-7, MCF-10A, MDA-MB-231, MDA-MB-468, HCC70, and SUM-159	[69]
Biosynthesized colloidal stabilized AgNPs	Spherical nanoparticles with average diameter 31.4 nm, negative zeta potential	Anticancer activity against MCF-7 and L-929 fibroblast tumor cell line due to the onset of redox stress	[70]

CONCLUSION

The bioconjugated AgNPs represent biogenic or biologically tolerant nanomaterials with diverse applications in molecular medicine, gene transfection, anticancer drug delivery, and bioimaging. The surface plasmonic characteristics of AgNPs coupled with the healing effect of biological conjugate offer multimodal applications with a single formulation. The strength of the bond of AgNPs with the conjugate molecule decides the cytotoxicity towards normal cells as a weaker bond sets free the anchored AgNPs that cause several detrimental effects in biological settings. A stronger bond, however, restrains the release of the anchored therapeutic that prevents the achievement of an optimal healing effect. Therefore, the utilization of bioconjugated AgNPs in biological applications must focus on the strength of the bond for conjugation and the ultimate fate of released AgNPs after achieving the deliberated application.

REFERENCES

[1]   Prasher, P.; Sharma, M.; Mehta, M.; Satija, S.; Aljabali, A.A.; Tambuwala, M.M.; Anand, K.; Sharma, N.; Dureja, H.; Jha, N.K.; Gupta, G.; Gulati, M.; Singh, S.K.; Chellappan, D.K.; Paudel, K.R.; Hansbro, P.M.; Dua, K. Current-status and applications of polysaccharides in drug delivery systems.

Colloid Interface Sci. Commun., **2021**, *42*: 100418.
[http://dx.doi.org/10.1016/j.colcom.2021.100418]

[2]   Fahmy, H.M.; Mosleh, A.M.; Elghany, A.A.; Shams-Eldin, E.; Serea, E.S.A.; Ali, S.A.; Shalan, A.E. Coated silver nanoparticles: synthesis, cytotoxicity, and optical properties. *RSC Advances,* **2019**, *9*(35), 20118-20136.
[http://dx.doi.org/10.1039/C9RA02907A]

[3]   Prasher, P.; Sharma, M.; Mudila, H.; Gupta, G.; Sharma, A.K.; Kumar, D.; Bakshi, H.A.; Negi, P.; Kapoor, D.N.; Chellappan, D.K.; Tambuwala, M.M.; Dua, K. Emerging trends in clinical implications of bio-conjugated silver nanoparticles in drug delivery. *Colloid Interface Sci. Commun.,* **2020**, *35*: 100244.
[http://dx.doi.org/10.1016/j.colcom.2020.100244]

[4]   Dolatabadi, A.; Noorbazargan, H.; Khayam, N.; Moulavi, P.; Zamani, N.; Asghari Lalami, Z.; Ashrafi, F. Ecofriendly Biomolecule-Capped *Bifidobacterium bifidum*-Manufactured Silver Nanoparticles and Efflux Pump Genes Expression Alteration in *Klebsiella pneumoniae. Microb. Drug Resist.,* **2021**, *27*(2), 247-257.
[http://dx.doi.org/10.1089/mdr.2019.0366] [PMID: 32635796]

[5]   Gopisetty, M.K.; Kovacs, D.; Igaz, N.; Ronavari, A.; Belteky, P.; Razga, Z.; Venglovecs, V.; Czoboz, B.; Boros, I.M.; Konya, Z.; Kiricsi, M. Endoplasmic reticulum stress: major player in size-dependent inhibition of P-glycoprotein by silver nanoparticles in multidrug-resistant breast cancer cells *J. Nanobiotechnol.,* **2019**, *17*, Article 9.

[6]   Srichaiyapol, O.; Thammawithan, S.; Siritongsuk, P.; Nasompag, S.; Daduang, S.; Klaynongsruang, S.; Kulchat, S.; Patramanon, R. Tannic Acid-Stabilized Silver Nanoparticles Used in Biomedical Application as an Effective Antimelioidosis and Prolonged Efflux Pump Inhibitor against Melioidosis Causative Pathogen. *Molecules,* **2021**, *26*(4), 1004.
[http://dx.doi.org/10.3390/molecules26041004] [PMID: 33672903]

[7]   Pollok, N.E.; Rabin, C.; Smith, L.; Crooks, R.M. Orientation-controlled bioconjugation of antibodies to silver nanoparticles. *Bioconjug. Chem.,* **2019**, *30*(12), 3078-3086.
[http://dx.doi.org/10.1021/acs.bioconjchem.9b00737] [PMID: 31730333]

[8]   Lee, S.H.; Jun, B-H. Silver nanoparticles: synthesis and application for nanomedicine. *Int. J. Mol. Sci.,* **2019**, *20*(4), 865.
[http://dx.doi.org/10.3390/ijms20040865] [PMID: 30781560]

[9]   Zhao, J.; Liu, P.; Ma, J.; Li, D.; Yang, H.; Chen, W.; Jiang, Y. Enhancement of Radiosensitization by Silver Nanoparticles Functionalized with Polyethylene Glycol and Aptamer As1411 for Glioma Irradiation Therapy. *Int. J. Nanomedicine,* **2019**, *14*, 9483-9496.
[http://dx.doi.org/10.2147/IJN.S224160] [PMID: 31819445]

[10]   Howard, D.; Sebastian, S.; Le, Q.V-C.; Thierry, B.; Kempson, I. Chemical Mechanisms of Nanoparticle Radiosensitization and Radioprotection: A Review of Structure-Function Relationships Influencing Reactive Oxygen Species. *Int. J. Mol. Sci.,* **2020**, *21*(2), 579.
[http://dx.doi.org/10.3390/ijms21020579] [PMID: 31963205]

[11]   Habiba, K.; Aziz, K.; Sanders, K.; Santiago, C.M.; Mahadevan, L.S.K.; Makarov, V.; Weiner, B.R.; Morell, G.; Krishnan, S. Enhancing Colorectal Cancer Radiation Therapy Efficacy using Silver Nanoprisms Decorated with Graphene as Radiosensitizers. *Sci. Rep.,* **2019**, *9*(1), 17120.
[http://dx.doi.org/10.1038/s41598-019-53706-0] [PMID: 31745177]

[12]   Sun, D.; Zhang, W.; Li, N.; Zhao, Z.; Mou, Z.; Yang, E.; Wang, W. Silver nanoparticles-quercetin conjugation to siRNA against drug-resistant Bacillus subtilis for effective gene silencing: *in vitro* and *in vivo. Mater. Sci. Eng. C,* **2016**, *63*, 522-534.
[http://dx.doi.org/10.1016/j.msec.2016.03.024] [PMID: 27040247]

[13]   Liu, Z.; Tan, H.; Zhang, X.; Chen, F.; Zhou, Z.; Hu, X.; Chang, S.; Liu, P.; Zhang, H. Enhancement of radiotherapy efficacy by silver nanoparticles in hypoxic glioma cells. *Artif. Cells Nanomed.*

Biotechnol., **2018**, *46*(sup3), S922-S930.
[http://dx.doi.org/10.1080/21691401.2018.1518912] [PMID: 30307330]

[14]   Yu, D.; Zhang, Y.; Lu, H.; Zhao, D. Silver nanoparticles coupled to anti-EGFR antibodies sensitize nasopharyngeal carcinoma cells to irradiation. *Mol. Med. Rep.,* **2017**, *16*(6), 9005-9010.
[http://dx.doi.org/10.3892/mmr.2017.7704] [PMID: 28990103]

[15]   Zhao, J.; Li, D.; Ma, J.; Yang, H.; Chen, W.; Cao, Y.; Liu, P. Increasing the accumulation of aptamer AS1411 and verapamil conjugated silver nanoparticles in tumor cells to enhance the radiosensitivity of glioma. *Nanotechnology,* **2021**, *32*(14): 145102.
[http://dx.doi.org/10.1088/1361-6528/abd20a] [PMID: 33296880]

[16]   Jyoti, K.; Singh, A.; Singh, T. Cytotoxic and radiosensitizing potential of silver nanoparticles against HepG-2 cells prepared by biosynthetic route using *Picrasma quassioides* leaf extract. *J. Drug. Disc. Sci. Technol.,* **2020**, *55*, Article 101479.

[17]   Fathy, M.M. Biosynthesis of Silver Nanoparticles Using Thymoquinone and Evaluation of Their Radio-Sensitizing Activity. *Bionanoscience,* **2020**, *10*(1), 260-266.
[http://dx.doi.org/10.1007/s12668-019-00702-3]

[18]   Wu, H.; Lin, J.; Liu, P.; Huang, Z.; Zhao, P.; Jin, H.; Ma, J.; Wen, L.; Gu, N. Reactive oxygen species acts as executor in radiation enhancement and autophagy inducing by AgNPs. *Biomaterials,* **2016**, *101*, 1-9.
[http://dx.doi.org/10.1016/j.biomaterials.2016.05.031] [PMID: 27254247]

[19]   Swanner, J.; Mims, J.; Carroll, D.L.; Akman, S.A.; Furdui, C.M.; Torti, S.V.; Singh, R.N. Differential cytotoxic and radiosensitizing effects of silver nanoparticles on triple-negative breast cancer and non-triple-negative breast cells. *Int. J. Nanomedicine,* **2015**, *10*, 3937-3953.
[PMID: 26185437]

[20]   Liu, P.; Huang, Z.; Chen, Z.; Xu, R.; Wu, H.; Zang, F.; Wang, C.; Gu, N. Silver nanoparticles: a novel radiation sensitizer for glioma? *Nanoscale,* **2013**, *5*(23), 11829-11836.
[http://dx.doi.org/10.1039/c3nr01351k] [PMID: 24126539]

[21]   Qiu, L.; Zhao, L.; Xing, C.; Zhan, Y. Redox-responsive polymer prodrug/ AgNPs hybrid nanoparticles for drug delivery. *Chin. Chem. Lett.,* **2018**, *29*(2), 301-304.
[http://dx.doi.org/10.1016/j.cclet.2017.09.048]

[22]   Nezami, S.; Sadeghi, M. pH-sensitive free AgNPs composite and nanocomposite beads based on starch as drug delivery systems. *Polym. Bull.,* **2020**, *77*(3), 1255-1279.
[http://dx.doi.org/10.1007/s00289-019-02801-3]

[23]   Gul, A.R.; Shaheen, F.; Rafique, R.; Bal, J.; Waseem, S.; Park, T.J. Grass-mediated biogenic synthesis of silver nanoparticles and their drug delivery evaluation: A biocompatible anti-cancer therapy. *Chem. Eng. J.,* **2020**.: 127202.

[24]   Benyettou, F.; Rezgui, R.; Ravaux, F.; Jaber, T.; Blumer, K.; Jouiad, M.; Motte, L.; Olsen, J.C.; Platas-Iglesias, C.; Magzoub, M.; Trabolsi, A. Synthesis of silver nanoparticles for the dual delivery of doxorubicin and alendronate to cancer cells. *J. Mater. Chem. B Mater. Biol. Med.,* **2015**, *3*(36), 7237-7245.
[http://dx.doi.org/10.1039/C5TB00994D] [PMID: 32262831]

[25]   Qiu, L.; Li, J-W.; Hong, C-Y.; Pan, C-Y. Silver Nanoparticles Covered with pH-Sensitive Camptothecin-Loaded Polymer Prodrugs: Switchable Fluorescence "Off" or "On" and Drug Delivery Dynamics in Living Cells. *ACS Appl. Mater. Interfaces,* **2017**, *9*(46), 40887-40897.
[http://dx.doi.org/10.1021/acsami.7b14070] [PMID: 29088537]

[26]   Bhanumathi, R.; Manivannan, M.; Thangaraj, R.; Kannan, S. Drug-Carrying Capacity and Anticancer Effect of the Folic Acid- and Berberine-Loaded Silver Nanomaterial To Regulate the AKT-ERK Pathway in Breast Cancer. *ACS Omega,* **2018**, *3*(7), 8317-8328.
[http://dx.doi.org/10.1021/acsomega.7b01347] [PMID: 30087941]

[27]   Fagbenro, K.A.O.; Saifullah, S.; Imran, M.; Parveen, S.; Rao, K.; Fasina, T.M.; Olasupo, I.A.; Adams, L.A.; Ali, I.; Shah, M.R. Egg proteins stabilized green silver nanoparticles as delivery system for hesperidin enhanced bactericidal potential against resistant *S. aureus*. *J. Drug Deliv. Sci. Technol.*, **2019**, *50*, 347-354.
[http://dx.doi.org/10.1016/j.jddst.2019.02.002]

[28]   Rozalen, M.; Polo, M.S.; Perales, M.F.; Widmann, T.J.; Utrilla, J.R. Synthesis of controlled-size silver nanoparticles for the administration of methotrexate drug and its activity in colon and lung cancer cells. *RSC Advances*, **2020**, *10*(18), 10646-10660.
[http://dx.doi.org/10.1039/C9RA08657A]

[29]   Li, Y.; Guo, M.; Lin, Z.; Zhao, M.; Xiao, M.; Wang, C.; Xu, T.; Chen, T.; Zhu, B. Polyethylenimine-functionalized silver nanoparticle-based co-delivery of paclitaxel to induce HepG2 cell apoptosis. *Int. J. Nanomedicine*, **2016**, *11*, 6693-6702.
[http://dx.doi.org/10.2147/IJN.S122666] [PMID: 27994465]

[30]   Kalindemirtas, F.D.; Kariper, A.; Hepokur, C.; Kuruca, S.E. Selective cytotoxicity of paclitaxel bonded silver nanoparticle on different cancer cells. *J. Drug Deliv. Sci. Technol.*, **2021**, *61*: 102265.
[http://dx.doi.org/10.1016/j.jddst.2020.102265]

[31]   Muhammad, N.; Zhao, H.; Song, W.; Gu, M.; Li, Q.; Liu, Y.; Li, C.; Wang, J.; Zhan, H. Silver nanoparticles functionalized Paclitaxel nanocrystals enhance overall anti-cancer effect on human cancer cells. *Nanotechnology*, **2021**, *32*(8): 085105.
[http://dx.doi.org/10.1088/1361-6528/abcacb] [PMID: 33197899]

[32]   Mohamed, N. Synthesis of Hybrid Chitosan Silver Nanoparticles Loaded with Doxorubicin with Promising Anti-cancer Activity. *Bionanoscience*, **2020**, *10*(3), 758-765.
[http://dx.doi.org/10.1007/s12668-020-00760-y]

[33]   Xu, L.; Wang, Y-Y.; Huang, J.; Chen, C-Y.; Wang, Z-X.; Xie, H. Silver nanoparticles: Synthesis, medical applications and biosafety. *Theranostics*, **2020**, *10*(20), 8996-9031.
[http://dx.doi.org/10.7150/thno.45413] [PMID: 32802176]

[34]   Kravets, V.; Almemar, Z.; Jiang, K.; Culhane, K.; Machado, R.; Hagen, G.; Kotko, A.; Dmytruk, I.; Spendier, K.; Pinchuk, A. Imaging of biological cells using luminescent silver nanoparticles. *Nanoscale Res. Lett.*, **2016**, *11*(1), 30.
[http://dx.doi.org/10.1186/s11671-016-1243-x] [PMID: 26781288]

[35]   Nie, F.; Ga, L.; Ai, J.; Wang, Y. PVP-templated novel fluorescent Ag-nanomaterial synthesis and its application to bioimaging. *Micro & Nano Lett.*, **2018**, *13*(6), 817-820.
[http://dx.doi.org/10.1049/mnl.2017.0871]

[36]   Mondal, I.; Raj, S.; Roy, P.; Poddar, R. Silver nanoparticles (AgNPs) as a contrast agent for imaging of animal tissue using swept-source optical coherence tomography (SSOCT). *Laser Phys.*, **2018**, *28*(1): 015601.
[http://dx.doi.org/10.1088/1555-6611/aa884b]

[37]   Kim, S.; Kim, S.; Lee, S.; Kwon, B.; Choi, J.; Hyun, J.W.; Kim, S. Characterization of the effects of silver nanoparticles on liver cell using HS-MAS NMR spectroscopy. *Bull. Korean Chem. Soc.*, **2011**, *32*, 2022-2026.

[38]   Singh, R.K.; Mishra, S.; Jena, S.; Panigrahi, B.; Das, B.; Jayabalan, R.; Parhi, P.K.; Mandal, D. Rapid colorimetric sensing of gadolinium by EGCG-derived AgNPs: the development of a nanohybrid bioimaging probe. *Chem. Commun. (Camb.)*, **2018**, *54*(32), 3981-3984.
[http://dx.doi.org/10.1039/C8CC01777H] [PMID: 29611570]

[39]   Yaraki, M.T.; Pan, Y.; Hu, F.; Yu, Y.; Liu, B.; Tan, Y.N. Nanosilver-enhanced AIE photosensitizer for simultaneous bioimaging and photodynamic therapy. *Mater. Chem. Front.*, **2020**, *4*(10), 3074-3085.
[http://dx.doi.org/10.1039/D0QM00469C]

[40]   Yang, F.; Wang, Q.; Gu, Z.; Fang, K.; Marriott, G.; Gu, N. Silver nanoparticle-embedded microbubble

as a dual-mode ultrasound and optical imaging probe. *ACS Appl. Mater. Interfaces,* **2013**, *5*(18), 9217-9223.
[http://dx.doi.org/10.1021/am4029747] [PMID: 23988030]

[41] Rasheed, W.; Shah, M.R.; Kazmi, M.H.; Shah, K.; Afridi, S. Sterically stabilized fluorescent silver nanoconjugates for optical discrimination of Cu(ii) in real samples and *in vitro* bioimaging of the Cu(ii) scavenging process in human lymphocytes by atomic force microscopy. *New J. Chem.,* **2016**, *40*(6), 5546-5554.
[http://dx.doi.org/10.1039/C6NJ00429F]

[42] Shahi, P.K.; Prakash, R.; Rai, S.B. Silver nanoparticles embedded hybrid organometallic complexes: Structural interactions, photo-induced energy transfer, plasmonic effect and optical thermometry. *AIP Adv.,* **2018**, *8*(6): 065117.
[http://dx.doi.org/10.1063/1.5020812]

[43] Jazayeri, S.D.; Ideris, A.; Shameli, K.; Moeini, H.; Omar, A.R. Gene expression profiles in primary duodenal chick cells following transfection with avian influenza virus H5 DNA plasmid encapsulated in silver nanoparticles. *Int. J. Nanomedicine,* **2013**, *8*, 781-790.
[http://dx.doi.org/10.2147/IJN.S39074] [PMID: 23459681]

[44] Tao, Y.; Ju, E.; Ren, J.; Qu, X. Metallization of plasmid DNA for efficient gene delivery. *Chem. Commun. (Camb.),* **2013**, *49*(84), 9791-9793.
[http://dx.doi.org/10.1039/c3cc45834b] [PMID: 24026136]

[45] Trimukhe, A.M.; Pofali, P.A.; Vaidya, A.A.; Koli, U.B.; Dandekar, P.; Deshmukh, R.R.; Jain, R.D. Pulsed plasma surface functionalized nanosilver for gene delivery. *Front. Biosci.,* **2020**, *25*(10), 1854-1874.
[http://dx.doi.org/10.2741/4881] [PMID: 32472761]

[46] Li, Y.; Lin, Z.; Xu, T.; Wang, C.; Zhao, M.; Xiao, M.; Wang, H.; Deng, N.; Zhu, B. Delivery of VP1 siRNA to inhibit the EV71 virus using functionalized silver nanoparticles through ROS-mediated signaling pathways. *RSC Advances,* **2017**, *7*(3), 1453-1463.
[http://dx.doi.org/10.1039/C6RA26472G]

[47] Peng, L-H.; Niu, J.; Zhang, C-Z.; Yu, W.; Wu, J-H.; Shan, Y.H.; Wang, X.R.; Shen, Y.Q.; Mao, Z-W.; Liang, W-Q.; Gao, J.Q. TAT conjugated cationic noble metal nanoparticles for gene delivery to epidermal stem cells. *Biomaterials,* **2014**, *35*(21), 5605-5618.
[http://dx.doi.org/10.1016/j.biomaterials.2014.03.062] [PMID: 24736021]

[48] Sarkar, K.; Banerjee, S.L.; Kundu, P.P.; Madras, G.; Chatterjee, K. Biofunctionalized surface-modified silver nanoparticles for gene delivery. *J. Mater. Chem. B Mater. Biol. Med.,* **2015**, *3*(26), 5266-5276.
[http://dx.doi.org/10.1039/C5TB00614G] [PMID: 32262602]

[49] John, M.S.; Nagoth, J.A.; Ramasamy, K.P.; Ballarini, P.; Mozzicafreddo, M.; Mancini, A.; Telatin, A.; Liò, P.; Giuli, G.; Natalello, A.; Miceli, C.; Pucciarelli, S. Horizontal gene transfer and silver nanoparticles production in a new *Marinomonas* strain isolated from the Antarctic psychrophilic ciliate *Euplotes focardii. Sci. Rep.,* **2020**, *10*(1), 10218.
[http://dx.doi.org/10.1038/s41598-020-66878-x] [PMID: 32576860]

[50] Lu, J.; Wang, Y.; Jin, M.; Yuan, Z.; Bond, P.; Guo, J. Both silver ions and silver nanoparticles facilitate the horizontal transfer of plasmid-mediated antibiotic resistance genes. *Water Res.,* **2020**, *169*: 115229.
[http://dx.doi.org/10.1016/j.watres.2019.115229] [PMID: 31783256]

[51] Zhao, K.; Rong, G.; Hao, Y.; Yu, L.; Kang, H.; Wang, X.; Wang, X.; Jin, Z.; Ren, Z.; Li, Z. IgA response and protection following nasal vaccination of chickens with Newcastle disease virus DNA vaccine nanoencapsulated with Ag@SiO$_2$ hollow nanoparticles. *Sci. Rep.,* **2016**, *6*(1), 25720.
[http://dx.doi.org/10.1038/srep25720] [PMID: 27170532]

[52] Casañas Pimentel, R.G.; Robles Botero, V.; San Martín Martínez, E.; Gómez García, C.; Hinestroza,

J.P. Soybean agglutinin-conjugated silver nanoparticles nanocarriers in the treatment of breast cancer cells. *J. Biomater. Sci. Polym. Ed.,* **2016**, *27*(3), 218-234.
[http://dx.doi.org/10.1080/09205063.2015.1116892] [PMID: 26540350]

[53]   Karuppaiah, A.; Siram, K.; Selvaraj, D.; Ramasamy, M.; Babu, D.; Sankar, V. Synergistic and enhanced anticancer effect of a facile surface modified non-cytotoxic silver nanoparticle conjugated with gemcitabine in metastatic breast cancer cells. *Mater. Today Commun.,* **2020**, *23*: 100884.
[http://dx.doi.org/10.1016/j.mtcomm.2019.100884]

[54]   Patra, N.; Kar, D.; Pal, A.; Behera, A. Antibacterial, anticancer, anti-diabetic and catalytic activity of bio-conjugated metal nanoparticles. *Adv. Natural Sci.: Nanosci. Nanotechnol.,* **2018**, *9*, Article 035001.

[55]   Vijayan, S.; Divya, K.; Jisha, M.S. *In vitro* anticancer evaluation of chitosan/biogenic silver nanoparticle conjugate on Si Ha and MDA MB cell lines. *Appl. Nanosci.,* **2020**, *10*(3), 715-728.
[http://dx.doi.org/10.1007/s13204-019-01151-w]

[56]   Majeed, S.; Aripin, F.H.B.; Shoeb, N.S.B.; Danish, M.; Ibrahim, M.N.M.; Hashim, R. Bioengineered silver nanoparticles capped with bovine serum albumin and its anticancer and apoptotic activity against breast, bone and intestinal colon cancer cell lines. *Mater. Sci. Eng. C,* **2019**, *102*, 254-263.
[http://dx.doi.org/10.1016/j.msec.2019.04.041] [PMID: 31146998]

[57]   Venkatesan, J.; Lee, J-Y.; Kang, D.S.; Anil, S.; Kim, S-K.; Shim, M.S.; Kim, D.G. Antimicrobial and anticancer activities of porous chitosan-alginate biosynthesized silver nanoparticles. *Int. J. Biol. Macromol.,* **2017**, *98*, 515-525.
[http://dx.doi.org/10.1016/j.ijbiomac.2017.01.120] [PMID: 28147234]

[58]   Baganizi, D.R.; Nyairo, E.; Duncan, S.A.; Singh, S.R.; Dennis, V.A. Interleukin-10 conjugation to carboxylated PVP-coated silver nanoparticles for improved stability and therapeutic efficacy. *Nanomater. (MDPI).,* **2017**, *7*, Article 165.

[59]   Capanema, N.S.V.; Carvalho, I.C.; Mansur, A.A.P.; Carvalho, S.M.; Lage, A.P.; Mansur, H.S. Hybrid Hydrogel Composed of Carboxymethylcellulose–Silver Nanoparticles–Doxorubicin for Anticancer and Antibacterial Therapies against Melanoma Skin Cancer Cells. *ACS Appl. Nano Mater.,* **2019**, *2*(11), 7393-7408. [MDPI].
[http://dx.doi.org/10.1021/acsanm.9b01924]

[60]   Karuppaiah, A.; Rajan, R.; Hariharan, S.; Balasubramaniam, D.K.; Gregory, M.; Sankar, V. Synthesis and Characterization of Folic Acid Conjugated Gemcitabine Tethered Silver Nanoparticles (FA-GE--AgNPs) for Targeted Delivery. *Curr. Pharm. Des.,* **2020**, *26*(26), 3141-3146.
[http://dx.doi.org/10.2174/1381612826666200316143239] [PMID: 32175835]

[61]   Yazdanparast, S.; Benvidi, A.; Banaei, M.; Nikukar, H.; Tezerjani, M.D.; Azimzadeh, M. Dual-aptamer based electrochemical sandwich biosensor for MCF-7 human breast cancer cells using silver nanoparticle labels and a poly(glutamic acid)/MWNT nanocomposite. *Microchim. Acta.,* **2018**, *185*, Article 405.

[62]   Sabela, M.I.; Makhanya, T.; Kanchi, S.; Shahbaaz, M.; Idress, D.; Bisetty, K. One-pot biosynthesis of silver nanoparticles using Iboza Riparia and Ilex Mitis for cytotoxicity on human embryonic kidney cells. *J. Photochem. Photobiol. B,* **2018**, *178*, 560-567.
[http://dx.doi.org/10.1016/j.jphotobiol.2017.12.010] [PMID: 29253815]

[63]   Austin, L.A.; Kang, B.; Yen, C-W.; El-Sayed, M.A. Nuclear targeted silver nanospheres perturb the cancer cell cycle differently than those of nanogold. *Bioconjug. Chem.,* **2011**, *22*(11), 2324-2331.
[http://dx.doi.org/10.1021/bc200386m] [PMID: 22010874]

[64]   Austin, L.A.; Ahmad, S.; Kang, B.; Rommel, K.R.; Mahmoud, M.; Peek, M.E.; El-Sayed, M.A. Cytotoxic effects of cytoplasmic-targeted and nuclear-targeted gold and silver nanoparticles in HSC-3 cells--a mechanistic study. *Toxicol. In Vitro,* **2015**, *29*(4), 694-705.
[http://dx.doi.org/10.1016/j.tiv.2014.11.003] [PMID: 25462594]

[65]   Datta, L.P.; Chatterjee, A.; Acharya, K.; De, P.; Das, M. Enzyme responsive nucleotide functionalized

silver nanoparticles with effective antimicrobial and anticancer activity. *New J. Chem.,* **2017**, *41*(4), 1538-1548.
[http://dx.doi.org/10.1039/C6NJ02955H]

[66]   Mussa Farkhani, S.; Asoudeh Fard, A.; Zakeri-Milani, P.; Shahbazi Mojarrad, J.; Valizadeh, H. Enhancing antitumor activity of silver nanoparticles by modification with cell-penetrating peptides. *Artif. Cells Nanomed. Biotechnol.,* **2017**, *45*(5), 1029-1035.
[http://dx.doi.org/10.1080/21691401.2016.1200059] [PMID: 27357085]

[67]   Zhu, B.; Li, Y.; Lin, Z.; Zhao, M.; Xu, T.; Wang, C.; Deng, N. Silver Nanoparticles Induce HePG-2 Cells Apoptosis Through ROS-Mediated Signaling Pathways. *Nanoscale Res. Lett.,* **2016**, *11*(1), 198.
[http://dx.doi.org/10.1186/s11671-016-1419-4] [PMID: 27075340]

[68]   Acharya, D.; Satapathy, S.; Somu, P.; Parida, U.K.; Mishra, G. Apoptotic Effect and Anticancer Activity of Biosynthesized Silver Nanoparticles from Marine Algae Chaetomorpha linum Extract Against Human Colon Cancer Cell HCT-116. *Biol. Trace Elem. Res.,* **2021**, *199*(5), 1812-1822.
[http://dx.doi.org/10.1007/s12011-020-02304-7] [PMID: 32743762]

[69]   Swanner, J.; Fahrenholtz, C.D.; Tenvooren, I.; Bernish, B.W.; Sears, J.J.; Hooker, A.; Furdui, C.M.; Alli, E.; Li, W.; Donati, G.L.; Cook, K.L.; Vidi, P-A.; Singh, R. Silver nanoparticles selectively treat triple-negative breast cancer cells without affecting non-malignant breast epithelial cells *in vitro* and *in vivo. FASEB Bioadv.,* **2019**, *1*(10), 639-660.
[http://dx.doi.org/10.1096/fba.2019-00021] [PMID: 32123812]

[70]   Khorrami, S.; Zarrabi, A.; Khaleghi, M.; Danaei, M.; Mozafari, M.R. Selective cytotoxicity of green synthesized silver nanoparticles against the MCF-7 tumor cell line and their enhanced antioxidant and antimicrobial properties. *Int. J. Nanomedicine,* **2018**, *13*, 8013-8024.
[http://dx.doi.org/10.2147/IJN.S189295] [PMID: 30568442]

CHAPTER 6

Silver Nanoparticles as Next Generation Antibiotics

Abstract: The oligodynamic effect of AgNPs, ability to generate reactive oxygen species, biofilm inhibition, appreciable biocidal effect, interaction and disruption of cell membranes, and DNA damage contribute towards the antibiotic properties of AgNPs. In addition, a remarkable synergistic effect with the representative antibiotics further supports the use of AgNPs as next generation antimicrobial agents. The evasion of the efflux pumps by the AgNPs pertaining to their high concentration at the membrane surface presents a significant feature for their application as future antimicrobial agents. The tethering of the customary antibiotics or antimicrobial therapeutics with AgNPs alleviates their cellular entry that further improves the inhibition potency and minimizes the therapeutic dose required to achieve an optimal relieving effect against the target pathogen. Mainly, the AgNPs disrupt redox balance on contacting the microbial cells that trigger a series of events, eventually causing the microbial cell death. The antimicrobial efficacy of AgNPs further led to the development of anti-fouling fabrics, antibacterial bandages, anti-sweat clothing, in addition to home appliances, such as washing devices and water purifiers. The present chapter succinctly highlights the stature of AgNPs as next generation antibiotics.

Keywords: Antibiotics, Biocidal effect, Microbicidal effect, Oligodynamic effect, Synergistic effect.

1. INTRODUCTION

Multiple factors contribute towards the biocidal properties of AgNPs. On coming in contact with the microbial cells, the fine-sized AgNPs aggregate at the cell membrane, which causes their internalization to the microbial cells by diffusion [1]. This property forms the basis of inhibition of the efflux pumps that offer antibiotic resistance to microbial cells. Notably, on entering the microbial cells, the AgNPs trigger oxidative stress, which marks the onset of various deleterious effects [2]. These effects include microbial DNA damage, ribosomal dysfunction, protein denaturation, mitochondrial dysfunction, and membrane lysis, which eventually causes microbial cell death. The AgNPs display a pronounced synergistic effect with contemporary antibiotics, which improves the efficacy of the latter [3]. Reportedly, the starch stabilized AgNPs displayed synergistic antifungal efficacy with ketoconazole and fluconazole, for the inhibition of BW-

Parteek Prasher & Mousmee Sharma
All rights reserved-© 2022 Bentham Science Publishers

P17 strain of candida albicans. The starch stabilized AgNPs reportedly disrupted the cell membrane integrity of the microbes by inhibiting the conversion of lanosterol to ergosterol by the lanosterol 14-α demethylase enzyme. As the enzyme holds responsibility for upholding the membrane integrity and elasticity, its inhibition results in the loss of membrane structure and function, thereby leading to microbial cell death [4]. Noticeably, the starch stabilized AgNPs bypassed the microbial efflux pumps, thereby debarring the membrane of its selective permeability that results in unrestricted access to cell organelles by the invading chemical agents and drugs. Importantly, in the presence of starch stabilized or cysteamine functionalized AgNPs, the microbial cells begin to show the clumping effect as an immediate response to the stress, which indicates the microbicidal effect of AgNPs. These observations confirmed the antibiofilm activity of the AgNPs, which causes microbial cell death. The considerable tolerance of polysaccharide capped AgNPs to the living cells further validates their efficacy in biomedical applications as next-generation antibiotics [5]. The non-dependence of the microbicidal activity of polysaccharide capped AgNPs on AcrAB-TolC efflux pump further validated the evasion of microbial efflux pumps by the AgNPs that leads to applications as future antibiotics. The inhibition of the efflux pumps occurs either by direct binding to the active site or by disrupting the efflux kinetics. The finer size of AgNPs causes an aggregation effect that leads to their rapid internalization to the microbial cells further explaining the achievement of an optimal biocidal effect irrespective of the efflux pump activity [6]. Kang *et al.* (2019) reported the hollow silver nanoparticles modified with tocopherol polyethylene glycol succinate for evading the microbial resistance to antibiotics. The hollow structure of the AgNP-based nanomaterial promoted the internalization of antibiotics to the microbial cells, thereby amplifying the biocidal effect [7].

The AgNP-based nanomaterial reportedly lowered the activity of AdeABC, and AdeIJK efflux pumps in drug resistant bacterial strains by a successful inhibition of the adeB, and adeJ efflux-pump genes. The silencing of these genes caused antibiotic accumulation on the microbial cell wall and its further entry to the cells improves the therapeutic efficacy. Jose *et al.* (2019) reported annihilation of drug resistant microbes in the presence of hybrid nanoparticles of silver in combination with silica. The antimicrobial efficacy of this nanosystem ameliorated in the presence of an efflux pump blocker Verapamil. The nanoparticles reportedly disrupted the structural integrity of the microbial cell wall, induced oxidative stress, and instigated the DNA damage that proved detrimental to the microbial cells. The complete annihilation of microbial cells occurred with the reported nanomaterial in the presence of a known efflux pump inhibitor, which suggested a synergistic effect. Importantly, the non-toxicity of the nanosystem further proved profitable in its utility in various biomedical applications. Fig. (**1**) indicates the biocidal effect of AgNPs on contacting the microbial cells [8].

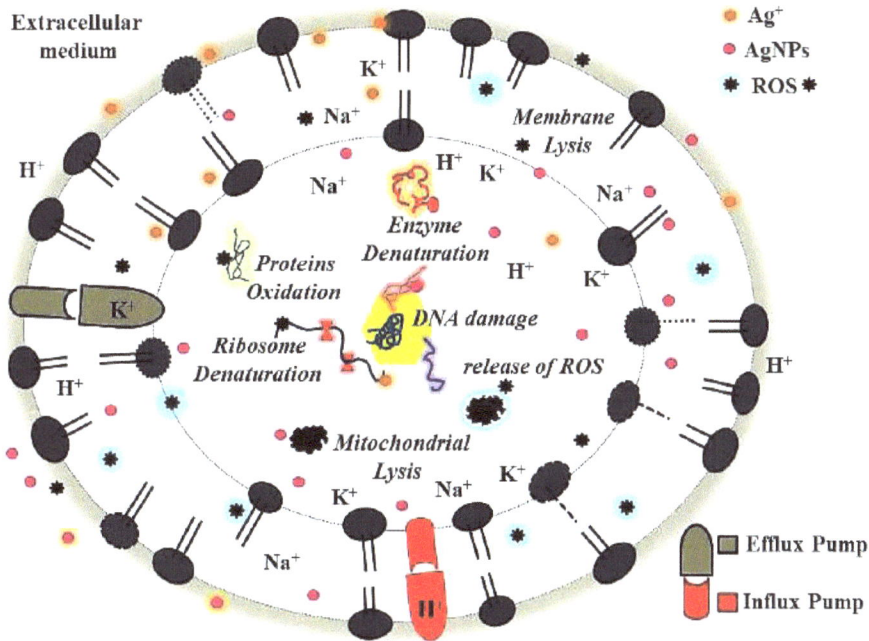

Fig. (1). Biocidal effect by AgNPs on contacting the microbial cell.

1.1. Microbial Membrane Disruption

Ahmed *et al.* (2018) explained the destruction of microbial membrane caused by the generation of the reactive oxygen species by AgNPs. The staining experiments on microbial isolates indicated the heightened production of ROS that deterred the membrane permeability and hampered the production of extracellular polymeric substances. The treatment of microbial cells of *Pseudomonas aeruginosa*, *E. coli*, and *Staphylococcus aureus* with AgNPs led to the development of gaps, pits, and fragmentation in the cell envelope. These pits and groves cause shrinkage of bacterial membrane, eventually causing the leakage of DNA and modifying the membrane permeability. Further investigations by SEM analysis confirmed structural irregularities in microbial cells on treatment with AgNPs, which include indentations, distortion, and development of irregular margins that caused bacterial cell death [9]. Su *et al.* (2009) reported the ROS-induced disruption of the cell membrane of *Staphylococcus aureus*, *Pseudomonas aeruginosa*, and *Streptococcus pyrogens* in the presence of AgNPs. The AgNPs with an average diameter of 30 nm reportedly adheres to the surface of bacterial cells, eventually causing a loss in the membrane integrity by the generation of ROS. AgNPs reportedly pass through the bacterial cell wall followed by the oxidation of surface proteins on the plasma membrane, which alters the homeostasis [10]. Bondarenko *et al.* (2018) reported that the plasma membrane of bacteria *E. coli* and *Pseudomonas aeruginosa* serves as the target for AgNPs for displaying the biocidal effect. Primarily, an optimal microbicidal effect by AgNPs requires a burst release of silver ions that ensures a rapid killing of bacterial cells and discourages the development of microbial resistance in target bacterial cells. The silver ions generated by AgNPs reportedly infiltrate the periplasm of the microbe *via* porins, and interact directly with the various components of plasma membrane, thereby causing its disruption [11].

1.2. Microbial Efflux Pump Modulation

Sanyasi *et al.* (2016) reported the role of polysaccharide-capped AgNPs in eliminating the multi-drug resistant bacteria *Bacillus subtilis*, *E. coli*, and *Salmonella typhimurium* by dismembering the bacterial cytoskeleton and inhibiting biofilm formation. The carboxy methyl-tamarind (CMT) polysaccharide coating on AgNPs imparted an inhibition of the bacterial cell division, in a dose dependent manner. In addition, the CMT coating on AgNPs caused bacterial membrane damage and prevented the formation of biofilms [12]. Behdad *et al.* (2020) reported the efflux-pump inhibition properties of AgNPs against the clinical isolates of *Acinetobacter baumannii* by the termination proton gradient that results in the disruption of membrane potential and a loss of proton motive force. The AgNPs reportedly bind directly to the active site of efflux pumps that discourages the extrusion of antibiotics from the bacterial cells. The competitive inhibition of AgNPs with antibiotics for the same binding site of the efflux pumps disrupts the efflux kinetics, thereby dismantling the microbial efflux [13]. Prasher *et al.* (2018) reported the starch stabilized AgNPs as modulators of the efflux pump in the BWP17 strain of *Candida albicans*. Reportedly, the nanoparticles competitively bind with the active site of efflux pump that caused a surge in the concentration of the substrate in the parent media. This confirmed direct interactions between the AgNPs and efflux pump, which led to the deactivation of the latter. The inhibition of microbial efflux played a significant role in ameliorating the antifungal efficacy of the drug ketoconazole. Investigations by Srichaiyapol *et al.* (2017) on the tannic acid stabilized AgNPs confirmed a prolonged inhibition of the microbial efflux at a sublethal concentration of nanoparticles in *Burkholderia pseudomallei*, the causative pathogen for Melioidosis. The nanoparticles proved highly effective in inhibiting the growth of ceftazidime-resistant strains. The evasion of microbial efflux pumps by surface functionalized AgNPs provided strong evidence for their applications as future antibiotics [14].

1.3. Anti-biofilm Activity

Anti-biofilm activity represents another mechanism adopted by AgNPs for displaying microbicidal effect. Kannan *et al.* (2021) reported the antibiofilm activity of biogenic AgNPs obtained from *Ludwigia octovalvis*. The AgNPs with an average diameter of 28-50 nm inhibited the biofilm formation in *E. coli*, *Staphylococcus aureus*, *Staphylococcus epidermidis*, *Proteus vulgaris* [15]. Jaiswal *et al.* (2017) reported the antibiofilm activity of curcumin-capped AgNPs with a diameter of 25-35 nm, where the curcumin fabrication resulted in a 20-fold improvement in the antibiofilm activity of AgNPs [16]. Faria *et al.* (2014) decorated graphene oxide with biogenic AgNPs for obtaining antibiofilm activity. The AgNPs reportedly distribute themselves homogeneously on the graphene oxide sheets with an average size of 8 nm. The graphene oxide sheets interact with the surface of bacterial cells *via* hydrogen bonding interactions occurring between the oxygenated groups on graphene oxide and the lipopolysaccharides of the microbial cell membrane. The hydrogen bonding facilitates the adhering of graphene oxide nanosheets on microbial cells that disrupts the transport of essential nutrients to the cell, eventually leading to their death [17]. Muthamil *et al.* (2018) reported the anti-biofilm activity of green synthesized AgNPs against *Candida* spp. The application of AgNPs caused a significant 80% reduction in the biofilm formation in the fungal species, higher than the commercial antifungal drugs at the same concentration. In addition, the AgNPs lowered the yeast-t--hyphal transition, the synthesis of exopolysaccharides, and the production of aspartyl proteinase that contribute towards the virulence of *Candida* spp [18]. Taglietti *et al.* (2014) anchored a monolayer of AgNPs onto amino-silanized glass surface for presenting the antibiofilm activity against *Staphylococcus epidermidis*. The prolonged release and high local concentration of silver ions led to the biofilm inhibition in the target microbe. The AgNPs displayed good stability in the aqueous media, which further supported their anti-biofilm application [19].

1.4. Microbial DNA Disruption

Abbas *et al.* (2019) reported the antibacterial efficacy of AgNPs by inducing DNA damage in the microbial cells of *Pseudomonas aeruginosa* and *Staphylococcus aureus* with a killing percentage of 81-95% after a direct exposure for 2h. The real time PCR analysis confirmed the damage inflicted on DNA by AgNPs that led to bacterial cell death [20]. Adeyemi *et al.* (2020) reported that the AgNPs inhibit the growth of Staphylococcus aureus, and E. coli by promoting oxidative stress mediated DNA damage in the bacterial cells. The AgNPs caused an elevated nitric oxide production, raised the level of lipid peroxidation, and caused the overexpression of superoxide dismutase. The alteration of redox status, infliction of DNA damage, and the activation of kynurenine pathway played a major role in the exhibition of antibacterial effect [21]. Butler *et al.* (2015) reported the correlation between the AgNPs particle size and their cellular uptake and genotoxicity in *Salmonella typhimurium* and *E. coli*. The test bacteria showed inefficiency in uptake the AgNPs with a diameter above 10 nm, due to limited diffusion and lack of facilitated uptake, such as endocytosis. However, the AgNPs with smaller size showed genotoxicity in bacterial cells due to direct interactions of AgNPs with DNA. The induction of genotoxicity also occurs by the generation of ROS, or through the shedding of ionic species [22]. Tamboli *et al.* (2013) provided a mechanistic approach for the AgNPs-mediated bacterial DNA disruption in *Salmonella typhimurium, Pseudomonas aeruginosa, E. coli, Exiguobacterium* sp. and *Staphylococcus aureus*. Scanning electron microscopy and gel electrophoresis confirmed the damage to the double stranded microbial DNA in the presence of spherical AgNPs with an average diameter of 30 nm. The bacterial extracellular enzymes supported the stabilization of AgNPs that enabled the achievement of an optimal inhibitory effect while maintaining the surface contact of AgNPs with bacterial cells [23].

1.5. Induction of Oxidative Stress

The colloidal stabilized AgNPs interacting with the bacterial cells generate Ag^+ ions that disturb the redox homeostasis in the microbial cells by triggering the release of free radicals [24]. Su *et al.* (2009) reported the generation of ROS in bacterial cells of *Staphylococcus aureus*, *Streptococcus pyrogens*, and *Pseudomonas aeruginosa* by the 40-50 nm sized AgNPs that caused a loss in the microbial membrane integrity pertaining to the release of ROS that deactivated the energy dependent metabolism in the microbial strain. The fabrication of AgNPs with thin silica plates offered surface for the local immobilization of AgNPs and prevented their entry to the cell membrane. The AgNPs reported in this study indicated lower cytotoxicity and offered high efficacy for combating drug resistant bacteria [25]. Qayyum *et al.* (2017) reported the obliteration of bacterial growth by the AgNPs that occurred due to the production of ROS by the nanoparticles in *Klebsiella Pneumoniae, E. coli*, and *Pseudomonas aeruginosa*. The AgNPs produced ROS on contacting the bacterial cells, followed by subsequent microbial membrane damage, and leakage of essential proteins from the cells. Similarly, ROS produced by AgNPs caused significant damage to the bacterial biofilm containing the essential polysaccharides and proteins required for the survival of bacterial colonies [26]. Huang *et al.* (2021) presented a rapid killing and capturing of bacteria *E. coli*, and *Staphylococcus aureus* for wound healing applications. The AgNPs generated Ag^+ ions and $^1O_2$ species in response to the near infrared irradiation that caused a significant 99.9% killing of bacteria.

The generation of redox stress in response to nanosilver and ionic silver seems to play a central role in determining the antimicrobial properties presented by the AgNPs [27] (Fig. **2**). illustrates the various approaches adopted by AgNPs for annihilating the microbial cells.

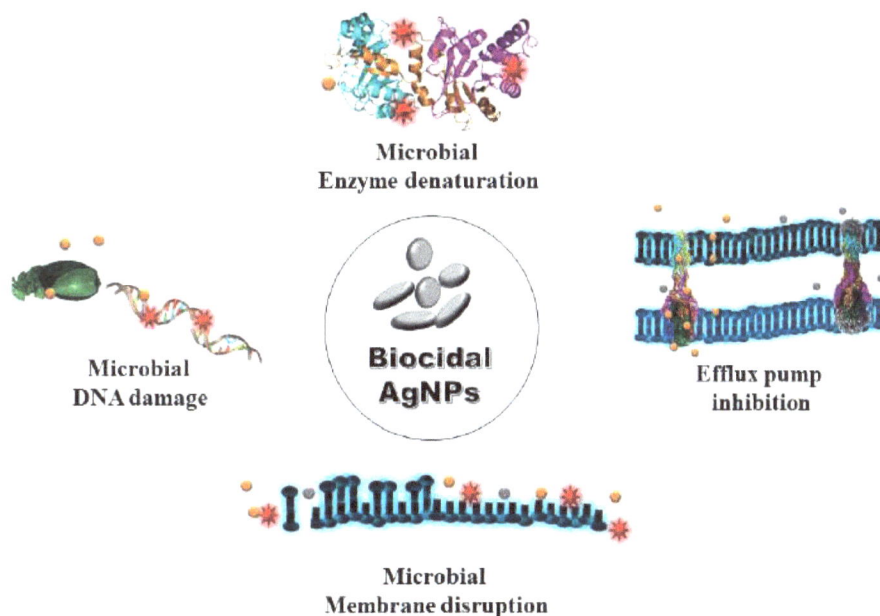

Fig. (2). Various approaches adopted by AgNPs for annihilating the microbial cells

1.6. Synergistic Effects

The AgNPs reportedly exhibit a synergistic effect with the β-lactam antibiotics in a concentration dependent manner. The hydroxy and amido groups in amoxicillin molecules react with nanosilver for chelate complexes, or interact *via* van der Waals interactions. This phenomenon promotes increased concentration of the antimicrobial on the microbe surface to achieve an optimal inhibition. Notably, the synergistic effect arises also due to the biocidal action of AgNPs. The nanoparticles approach the hydrophobic microbial membrane more effectively due to their hydrophilic nature, thereby promoting the cellular entry of the carried drug molecules [28] (Li *et al.* 2005). Fayaz *et al.* (2010) reported the synergistic effect of AgNPs with antibiotics in the inhibition of gram-positive and gram-negative bacteria. Ampicillin displayed the highest synergistic effect among kanamycin, erythromycin, and chloramphenicol. The antibiotic-AgNPs complex reacts with microbial DNA to prevent its unwinding, thereby causing serious damage to the microbial cells. The reaction of antibiotics such as ampicillin with the microbial cell wall created perforations in the membrane that facilitated the cellular internalization of antibiotic-AgNPs complex [29]. Deng *et al.* (2016) investigated the mechanistic insights for the synergistic effect of AgNPs with antibiotics. The antibiotic molecules belonging to the tetracycline class showed a direct binding to the bacterial cells, thereby releasing the attached AgNPs that resulted in an enhanced local concentration of silver ions at the bacterial cell surface. The presence of silver ion proves toxic to the microbial cells. The silver ions onset a series of events on their cellular entry to the microbial cells, including the onset of oxidative stress, generation of reactive oxygen species, and interfering with essential cellular processes that eventually cause the microbial cell death [30]. Jia *et al.* (2012) reported the synergistic antimicrobial effects of polyaniline with AgNPs. Mainly, the interactions between AgNPs and polyaniline created redox imbalance in the target microbial cells, thereby resulting in their death. The AgNPs showed lower interactions with gram-positive bacteria; while the gram-negative bacteria with a low-osmotic pressure multi-layered cell wall structure facilitates the interactions with AgNPs [31]. Carrizales *et al.* (2018) reported the *in vitro* synergism of antibiotics with AgNPs for the treatment of multi-resistant uropathogens. The investigations on 12 microbial strains indicated that ampicillin on combining with AgNPs displayed one synergy, four partial synergies, and additive effect in four strains of microbes. Amikacin in combination with AgNPs displayed synergistic and partial synergistic effect on three and eight microbial strains, with additive effects in one strain. Notably, the combination of AgNPs with antibiotics displayed only trivial cytotoxicity that indicated their applications as next generation antibiotics [32]. Smekalova *et al.* (2016) investigated the antimicrobial effect of various antibiotics in combination with AgNPs. The combination significantly improved the IC50 of the antibiotics and enhanced the sensitivity of some strains towards the antibiotic dose. The gram-positive bacteria, however, showed lesser susceptibility towards the synergistic combination, due to the thicker peptidoglycan layer present on the cell wall. Mainly, gentamycin exhibited a pronounced synergistic effect. The AgNPs reportedly enhanced the antimicrobial activity of gentamycin, although the strongest synergistic effect appeared for penicillin G [33]. McShan *et al.* (2015) reported the synergistic antimicrobial effect of AgNPs in combination with ineffective antibiotics. The antibiotics neomycin and tetracycline deliver antibiotic effect by binding with the bacterial membrane proteins. Therefore, their combination with AgNPs assists in cellular binding to the AgNPs, which on cellular entry result in the generation of oxidative stress for killing the microbial cells. Similarly, penicillin binds to the transpeptidase protein by covalent bonds. Therefore, the antibiotic does not bind effectively with AgNPs, resulting in the achievement of lower synergistic effect [34]. Devi *et al.* (2012) reported the antimicrobial synergistic effect of AgNPs against *Staphylococcus aureus, Streptococcus pyrogenes, Salmonella enterica, and Enterococcus faecalis*, in combination with erythromycin, chloramphenicol, ciprofloxacin, and methicillin. The mycosynthesis of green AgNPs resulted in a particle size of 5-50 nm, which demonstrated cellular internalization to show the biocidal effect. The AgNPs in combination with antibiotics resulted in the increase in the inhibition zone against the test bacterium, compared to the antibiotics alone, which confirmed a synergistic effect [35]. Panacek *et al.* (2016) demonstrated the synergistic antibacterial efficacy of AgNP + antibiotic combination at lower concentrations. The presence of AgNPs considerably lowered the MIC of antibiotics for inhibiting the test bacterial strains. The presence of AgNPs restored the vulnerability of *Escherichia coli* towards ampicillin, which indicated that the nanoparticles pose interference with the bacterial efflux system. Notably, the insignificant cytotoxicity of nanoparticles towards mammalian cells support their further development as impending antibiotic formulations [36]. Naqvi *et al.* (2013) reported the synergistic effect of AgNPs against drug-resistant bacteria *Staphylococcus aureus, Fusarium semitactum, and Aspergillus flavus.* The AgNPs reportedly form a complex with antibiotics that result in the disruption of the peptidoglycan layer in bacteria. The presence of positive charge on nanoparticles mainly causes an attack on the oppositely charged transmembrane protein, which eventually causes deformations in cell membrane and blocks the transport channels. The efficient interactions of AgNPs with thiol groups of peptides, and phosphate groups in the microbial DNA serve as the basis of the enhanced antimicrobial effect of antibiotics in combination with AgNPs. The local aggregation of antibiotic molecules complexed with AgNPs at the microbial cell surface further augments the bacterial lysis. Overall, the AgNPs caused a 2.7-fold increase in the bactericidal efficacy of the antibiotics, as indicated by an increase in the zone of inhibition in the *in vitro* antibacterial analysis [37]. Mazur *et al.* (2020) presented the synergistic ROS-mediated biocidal effect of AgNPs in combination with gentamycin against the multidrug-resistant strains of *Staphylococcus epidermidis*. The combining of nanoparticles lowered the MIC of gentamycin by 16-times, mainly due to the release of the reactive oxygen species [38]. Ghosh *et al.* (2013) presented the synergistic antibacterial effect of AgNPs in combination with cinnamaldehyde against the spore-forming bacteria. The bacteria kill curve suggested a rapid bactericidal effect caused by the combination of AgNPs with cinnamaldehyde, causing extensive damage to the bacterial cells without displaying toxicity. The hemolysis assay confirmed that the combination produced negligible cytotoxicity towards the human cells [39]. Habash *et al.* (2014) reported the synergism of AgNPs with aztreonam for the inhibition of *Pseudomonas aeruginosa* PAO1 biofilms. Several pathogenic microbes form biofilms to exhibit resistance towards the antibiotics, which prevents the penetration and dispersion of the drug in microbial cell. The AgNPs with particle size 10 nm displayed the highest synergism with aztreonam by improving the infiltration of the aztreonam in the biofilm matrix. This phenomenon promoted optimal antibacterial effect by the antibiotic in the presence of AgNPs [40]. Krychowiak *et al.* (2018) reported the synergistic effect of AgNPs with naphthoquinones for the inhibition of *Staphylococcus aureus*. The release of ionic silver caused membrane damage in the bacterial cell wall, which showed disruptions in treatment with this synergistic combination. The bacterial cells showed weak aggregation in the presence of AgNP-naphthoquinone that indicated the possible inhibition of the formation of biofilm or colonization of bacterial cells. The synergistic effect also appeared for the bacterial strains resistant towards antibiotics. Importantly, the presence of AgNPs lowered the effective dose of naphthoquinone required for the inhibition of target bacterial cells. This lowered dose further minimized the cytotoxicity of naphthoquinone on healthy cells. Table **1** presents the synergistic effect of AgNPs with various antibiotics [41].

Table 1. Synergistic effect of AgNPs with antibiotics and biocidal agents on various microbes.

Morphology of AgNPs	Antibiotic/ Microbicide	Synergistic Effect Against Pathogen	Refs.
Spherical, 1-100 nm	Oxacillin, Neomycin	*Staphylococcus aureus, E. coli, Salmonella spp., Candida albicans*	[42]
Polydispersed, 71-201 nm	Ofloxacin, Ciprofloxacin, Gentamycin, Ampicillin	*Vibrio parahaemolyticus, Aeromonas hydrophila, Edwardsiella tarda*	[43]
Spherical, 15-20 nm	vancomycin, oleandomycin, ceftazidime, rifampicin, penicillin G, neomycin, cephazolin, novobiocin, carbenicillin, lincomycin, tetracycline, and erythromycin	*Pseudomonas aeruginosa, Escherichia coli, Staphylococcus aureus, Candida albicans, Bacillus subtilis*	[44]
Spherical, 35 nm	Chloramphenicol, Kanamycin, Ampicillin, Aztreonam, Biapenem	*Escherichia coli, Salmonella Enterica serovar, Staphylococcus aureus, Bacillus subtilis*	[45]
Spherical, 26 nm	Cefuroxine, Gentamycin	*Staphylococcus aureus, Methicillin Resistant Staphylococcus aureus, Streptococcus mutans, Streptococcus oralis, Streptococcus gordonii, Enterococcus faecalis, Escherichia coli, Pseudomonas aeruginosa*	[46]
Spherical, 80-120 nm	Cefixime, Amoxicillin, Levofloxacin	*Pseudomonas aeruginosa, Salmonella typhi, Klebsiella pneumoniae, E. coli*	[47]
Spherical, 20 nm	Gentamycin, Chloramphenicol	*Enterococcus faecalis, Klebsiella pneumoniae*	[48]

(Table 1) cont.....

Morphology of AgNPs	Antibiotic/ Microbicide	Synergistic Effect Against Pathogen	Refs.
Spherical, 35-60 nm	Ciprofloxacin	*Staphylococcus aureus, Pseudomonas aeruginosa*	[49]
Spherical, 20-170 nm	Ampicillin	*Staphylococcus aureus, E. coli*	[50]
Variable	Kanamycin, Doxycycline, Tetracycline, Cefoxitin, Oxacillin	*E. coli, Staphylococcus aureus, Staphylococcus epidermidis*	[51]
Spherical, 10-30 nm	Gentamycin	*Staphylococcus aureus*	[52]
Spherical, 5-15 nm	Amoxycillin, Ampicillin, Erythromycin, Kanamycin, Tetracycline,	*Bacillus cereus, Staphylococcus aureus, Pseudomonas aeruginosa, Klebsiella, pneumoniae, Salmonella typhimurium, Vibrio vulnificus*	[53]
Spherical, variable size	Polymyxin B	*Pseudomonas aeruginosa*	[54]
Spherical, 35.50 nm	Amoxicillin	*E. coli*	[55]
Spherical 80-120 nm	Levofloxacin, Amoxicillin, Cefixime	*E. coli, Klebsiella pneumoniae, Staphylococcus aureus, Salmonella typhi, Pseudomonas aeruginosa*	[47]
Spherical 70-80 nm	Violacein	*E. coli, Staphylococcus aureus*	[56]
Spherical <20 nm	Ketoconazole	*Candida albicans*	[57]
Spherical 5 nm	Tobramycin	*E. coli*	[58]
Spherical 10-100 nm	Cephalexin	*Staphylococcus aureus*	[59]
Spherical 10-12 nm	Tetracycline, Ampicillin	*E. coli*	[60]
Spherical 1-10 nm	Azithromycin, Levofloxacin, Tetracycline	*E. coli, Klebsiella, pneumoniae, Staphylococcus aureus, Enterococcus faecalis*	[61]
Spherical 13.5-25.8 nm	Amoxicillin	*Streptococcus mutans, Staphylococcus aureus*	[62]
Spherical 13 nm	Epoxiconazole	*Setosphaeria turcica*	[63]
Spherical 8-21 nm	Ciprofloxacin, Methicillin, Gentamycin, Rifampicin	*Staphylococcus epidermidis* and *Staphylococcus haemolyticus*	[64]
Spherical 20-80 nm	Streptomycin, Rifampicin, Chloramphenicol, Novobiocin, Ampicillin	*Escherichia coli, Pseudomonas aeruginosa,* and *Staphylococcus aureus*	[65]
Spherical 5-12 nm	Carbapenem, Polymixin B, Rifampicin	*Acinetobacter baumannii*	[66]
Spherical 28 nm	Carbapenem, β-lactam	*Enterobacteriaceae*	[67]
Spherical 20-30 nm	Kanamycin, Cefepime, Amikacin, Ampicillin, Cefetaxime	*Bacillus cereus, Staphylococcus aureus, E. coli, Bacillus subtilis, Staphylococcus epidermidis, Salmonella typhimurium, Serratia marcescens*	[68]

(Table 1) cont.....

Morphology of AgNPs	Antibiotic/ Microbicide	Synergistic Effect Against Pathogen	Refs.
Crystalline 16.4-29.72 nm	Tobramycin, Ciprofloxacin, Vancomycin, Gentamycin, Cloxacillin	*Staphylococcus aureus, E. coli, Klebsiella pneumoniae, Pseudomonas aeruginosa*	[69]
Spherical 270 nm	Vancomycin	*Staphylococcus aureus, E. coli*	[70]
Spherical 10.6 nm	Gentamicin	*Staphylococcus aureus*	[71]
Spherical 121 nm	Chloramphenicol, Kanamycin, Vancomycin, Chloramphenicol, Polymyxin B	*Pseudomonas aeruginosa*	[72]
Spherical 5-21 nm	Vancomycin	*Staphylococcus aureus, Pseudomonas aeruginosa, Streptococcus pneumoniae*	[73]
Spherical 14-42 nm	Gentamycin, Ciprofloxacin, Penicillin, Chloramphenicol, Fusidic acid, Erythromycin	*Salmonella typhi, Vibrio cholera, Staphylococcus aureus*	[74]
Spherical 5.8 nm	Gentamycin, streptomycin, ciprofloxacin	*Staphylococcus aureus, Pseudomonas aeruginosa, E. coli*	[75]
Spherical 26 nm	Erythromycin, Chloramphenicol, Ampicillin, Cefuroxime, Cefpodoxime	*Streptococcus sp., E. coli, Enterococcus faecalis, Staphylococcus aureus*	[46]
Spherical 1-100 nm	Ampicillin, Streptomycin, Gentamycin,	*E. coli, Salmonella typhimurium, Pseudomonas aeruginosa*	[76]
Spherical 15 nm	Gentamycin, Chloramphenicol	*E. coli, Bacillus cereus, Staphylococcus aureus, Proteus mirabilis*	[77]
Spherical 8-12 nm	Chloramphenicol, Tetracycline, Doxycycline, Amoxicillin, Kanamycin, Gentamycin, Amikacin, Ampicillin, Trimethoprim, Penicillin, Ceftazidime	*Pseudomonas aeruginosa, Shigella sonnie, Salmonella typhimurium, Staphylococcus mutans, Enterobacter aerogenes, E. coli, Acinetobacter baumannii. Streptococcus mutans*	[78]
Spherical 10-90 nm	Ampicillin, Polymyxin, Chloramphenicol, Gentamycin, Cephalothin, Amoxyclav, Cefpirome, Clotrimazole, Amikacin, Tetracycline, Penicillin	*Staphylococcus aureus, Bacillus cereus, Candida glabrata, Candida albicans, Cryptococcae neoformans*	[79]
Spherical 5-40 nm	Streptomycin, Vancomycin, Tetracycline, Gentamycin, Erythromycin, Ciprofloxacin	*E. coli, Staphylococcus aureus*	[80]
Spherical 15-20 nm	Amoxicillin, Vancomycin, Clarithromycin, Linezolid, Azithromycin	*Methicillin resistant Staphylococcus aureus (MRSA)*	[81]

(Table 1) cont.....

Morphology of AgNPs	Antibiotic/ Microbicide	Synergistic Effect Against Pathogen	Refs.
Spherical 30-100 nm	Novobiocin, Vancomycin, Rifampicin, Penicillin G, Oleandomycin, Lincomycin, Cycloheximide	*Bacillus anthracis, Staphylococcus aureus, Salmonella enterica, Candida albicans, Bacillus cereus, Vibrio parahaemolyticus*	[82]
Spherical 9-15 nm	Tobramycin, Chloramphenicol,	*Enterobacter aerogenes, Citrobacter freundii, Klebsiella pneumoniae, Hafnia alvei, Salmonella typhii, Proteus mirabilis, E. coli*	[83]
Spherical 5-50 nm	Kanamycin, Tetracycline	*Bacillus subtilis, Staphylococcus aureus, E. coli, Candida albicans, Malassezia furfur*	[84]

CONCLUSION

The antimicrobial properties of AgNPs arise due to their unique ability to generate redox stress in the microbial cells by the production of ROS. These species trigger a cascade of events, including membrane disruption, DNA damage, mitochondrial dysfunction, ribosomal damage, cytoplasm leakage that eventually cause microbial cell lysis. AgNPs also show synergistic inhibition of microbial growth when used in combination with representative antibiotics. The contemporary antibiotics suffer an efflux from the microbial cells due to the activity of membrane bound efflux pumps. The AgNPs, however, evade the active microbial efflux by maintaining an unusually high concentration at the surface of microbial membrane that allows their internalization to the target cells *via* a concentration gradient. The bioconjugation of antibiotics with AgNPs improves their cellular internalization for displaying an optimal inhibitory effect. However, it necessitates the appraisal of toxicity profile of AgNPs before using the direct applications of nanosilver formulations as antimicrobial agents. Similarly, the immunogenicity of AgNPs poses a major concern in their clinical success as future antimicrobials. Hence, the impending antimicrobial development programs must take into consideration the deleterious effects of AgNPs to ensure a high clinical success rate.

REFERENCES

[1]   Mikhailova, E.O. Silver nanoparticles: Mechanism of action and probable bio-application. *J. Funct. Biomater.,* **2020**, *11*(4), 84.
[http://dx.doi.org/10.3390/jfb11040084] [PMID: 33255874]

[2]   Salleh, A.; Naomi, R.; Utami, N.D.; Mohammad, A.W.; Mahmoudi, E.; Mustafa, N.; Fauzi, M.B. The potential of silver nanoparticles for antiviral and antibacterial applications: A mechanism of action. *Nanometer. (MDPI Basel).,* **2020**, *10*, Article 1566.

[3]   Prasher, P.; Singh, M.; Mudila, H. Silver nanoparticles as antimicrobial therapeutics: current

perspectives and future challenges. *3 Biotech.,* **2018**, *8*, Article 411.

[4]   Prasher, P.; Singh, M.; Mudila, H. Green synthesis of silver nanoparticles and their antifungal properties. *Bionanoscience,* **2018**, *8*(1), 254-263.
[http://dx.doi.org/10.1007/s12668-017-0481-4]

[5]   Singh, M.; Prasher, P. Ultrafine silver nanoparticles: Synthesis and biocidal studies. *Bionanoscience,* **2018**, *8*(3), 735-741.
[http://dx.doi.org/10.1007/s12668-018-0522-7]

[6]   Mishra, M.; Kumar, S.; Majhi, R.K.; Goswami, L.; Goswami, C.; Mohapatra, H. Antibacterial efficacy of polysaccharide capped silver nanoparticles is not compromised by AcrAB-TolC efflux pump. *Front. Microbiol.,* **2018**, *9*, 823.
[http://dx.doi.org/10.3389/fmicb.2018.00823] [PMID: 29780364]

[7]   Kang, X-Q.; Qiao, Y.; Lu, X-Y.; Jiang, S-P.; Li, W-S.; Wang, X-J.; Xu, X-L.; Qi, J.; Xiao, Y.H.; Du, Y.Z. Tocopherol polyethylene glycol succinate-modified hollow silver nanoparticles for combating bacteria-resistance. *Biomater. Sci.,* **2019**, *7*(6), 2520-2532.
[http://dx.doi.org/10.1039/C9BM00343F] [PMID: 30968093]

[8]   Jose, J.; Anas, A.; Jose, B.; Puthirath, A.B.; Athiyanathil, S.; Jasmin, C.; Anantharaman, M.R.; Nair, S.; Subrahmanyam, C.; Biju, V. Extinction of antimicrobial resistant pathogens using silver embedded silica nanoparticles and an efflux pump blocker. *ACS Appl. Bio Mater.,* **2019**, *2*(11), 4681-4686.
[http://dx.doi.org/10.1021/acsabm.9b00614] [PMID: 35021465]

[9]   Ahmed, B.; Hashmi, A.; Khan, M.S.; Musarrat, J. ROS mediated destruction of cell membrane, growth and biofilms of human bacterial pathogens by stable metallic AgNPs functionalized from bell pepper extract and quercetin. *Adv. Powder Technol.,* **2018**, *29*(7), 1601-1616.
[http://dx.doi.org/10.1016/j.apt.2018.03.025]

[10]   Su, H-L.; Chou, C-C.; Hung, D-J.; Lin, S-H.; Pao, I-C.; Lin, J-H.; Huang, F-L.; Dong, R-X.; Lin, J-J. The disruption of bacterial membrane integrity through ROS generation induced by nanohybrids of silver and clay. *Biomaterials,* **2009**, *30*(30), 5979-5987.
[http://dx.doi.org/10.1016/j.biomaterials.2009.07.030] [PMID: 19656561]

[11]   Bondarenko, O.M.; Sihtmäe, M.; Kuzmičiova, J.; Ragelienė, L.; Kahru, A.; Daugelavičius, R. Plasma membrane is the target of rapid antibacterial action of silver nanoparticles in *Escherichia coli* and *Pseudomonas aeruginosa. Int. J. Nanomedicine,* **2018**, *13*, 6779-6790.
[http://dx.doi.org/10.2147/IJN.S177163] [PMID: 30498344]

[12]   Sanyasi, S.; Majhi, R.K.; Kumar, S.; Mishra, M.; Ghosh, A.; Suar, M.; Satyam, P.V.; Mohapatra, H.; Goswami, C.; Goswami, L. Polysaccharide-capped silver Nanoparticles inhibit biofilm formation and eliminate multi-drug-resistant bacteria by disrupting bacterial cytoskeleton with reduced cytotoxicity towards mammalian cells. *Sci. Rep.,* **2016**, *6*(1), 24929.
[http://dx.doi.org/10.1038/srep24929] [PMID: 27125749]

[13]   Behdad, R.; Pargol, M.; Mirzaie, A.; Karizi, S.Z.; Noorbazargan, H.; Akbarzadeh, I. Efflux pump inhibitory activity of biologically synthesized silver nanoparticles against multidrug-resistant *Acinetobacter baumannii* clinical isolates. *J. Basic Microbiol.,* **2020**, *60*(6), 494-507.
[http://dx.doi.org/10.1002/jobm.201900712] [PMID: 32301139]

[14]   Srichaiyapol, O.; Thammawithan, S.; Siritongsuk, P.; Nasompag, S.; Daduang, S.; Klaynongsruang, S.; Kulchat, S.; Patramanon, R. Tannic Acid-Stabilized Silver Nanoparticles Used in Biomedical Application as an Effective Antimelioidosis and Prolonged Efflux Pump Inhibitor against Melioidosis Causative Pathogen. *Molecules,* **2021**, *26*(4), 1004.
[http://dx.doi.org/10.3390/molecules26041004] [PMID: 33672903]

[15]   Kannan, D.S.; Mahboob, S.; Al-Ghanim, K.A.; Venkatachalam, P. Antibacterial, Antibiofilm and Photocatalytic Activities of Biogenic Silver Nanoparticles from *Ludwigia octovalvis. J. Cluster Sci.,* **2021**, *32*(2), 255-264.
[http://dx.doi.org/10.1007/s10876-020-01784-w]

[16]   Jaiswal, S.; Mishra, P. Antimicrobial and antibiofilm activity of curcumin-silver nanoparticles with improved stability and selective toxicity to bacteria over mammalian cells. *Med. Microbiol. Immunol. (Berl.),* **2018**, *207*(1), 39-53.
[http://dx.doi.org/10.1007/s00430-017-0525-y] [PMID: 29081001]

[17]   Faria, A.F.; Moraes, A.C.M.; Marcato, P.D.; Stefani, D.; Martinez, T.; Duran, N.; Filho, A.G.S.; Brandelli, A.; Alves, O.L. Eco-friendly decoration of graphene oxide with biogenic silver nanoparticles: antibacterial and antibiofilm activity. *J. Nanopart. Res.,* **2014**, *16*(2), 2110.
[http://dx.doi.org/10.1007/s11051-013-2110-7]

[18]   Muthamil, S.; Devi, V.A.; Balasubramaniam, B.; Balamurugan, K.; Pandian, S.K. Green synthesized silver nanoparticles demonstrating enhanced *in vitro* and *in vivo* antibiofilm activity against Candida spp. *J. Basic Microbiol.,* **2018**, *58*(4), 343-357.
[http://dx.doi.org/10.1002/jobm.201700529] [PMID: 29411881]

[19]   Taglietti, A.; Arciola, C.R.; D'Agostino, A.; Dacarro, G.; Montanaro, L.; Campoccia, D.; Cucca, L.; Vercellino, M.; Poggi, A.; Pallavicini, P.; Visai, L. Antibiofilm activity of a monolayer of silver nanoparticles anchored to an amino-silanized glass surface. *Biomaterials,* **2014**, *35*(6), 1779-1788.
[http://dx.doi.org/10.1016/j.biomaterials.2013.11.047] [PMID: 24315574]

[20]   Abbas, W.S.; Atwan, Z.W.; Abdulhussein, Z.R.; Mahdi, M.A. Preparation of silver nanoparticles as antibacterial agents through DNA damage. *Mater. Technol.,* **2019**, *34*(14), 867-879.
[http://dx.doi.org/10.1080/10667857.2019.1639005]

[21]   Adeyemi, O.S.; Shittu, E.O.; Akpor, O.B.; Rotimi, D.; Batiha, G.E. Silver nanoparticles restrict microbial growth by promoting oxidative stress and DNA damage. *EXCLI J.,* **2020**, *19*, 492-500.
[PMID: 32398973]

[22]   Butler, K.S.; Peeler, D.J.; Casey, B.J.; Dair, B.J.; Elespuru, R.K. Silver nanoparticles: correlating nanoparticle size and cellular uptake with genotoxicity. *Mutagenesis,* **2015**, *30*(4), 577-591.
[http://dx.doi.org/10.1093/mutage/gev020] [PMID: 25964273]

[23]   Tamboli, D.P.; Lee, D.S. Mechanistic antimicrobial approach of extracellularly synthesized silver nanoparticles against gram positive and gram negative bacteria. *J. Hazard. Mater.,* **2013**, *260*, 878-884.
[http://dx.doi.org/10.1016/j.jhazmat.2013.06.003] [PMID: 23867968]

[24]   Dakal, T.C.; Kumar, A.; Majumdar, R.S.; Yadav, V. Mechanistic Basis of Antimicrobial Actions of Silver Nanoparticles. *Front. Microbiol.,* **2016**, *7*, 1831.
[http://dx.doi.org/10.3389/fmicb.2016.01831] [PMID: 27899918]

[25]   Platania, V.; Kerou, A.K.; Karamanidou, T.; Kouki, M.; Tsouknidas, A.; Chatzinikolaidoi, M. **2022**.*Antibacterial effect of colloidal suspensions varying in silver nanoparticles and ion concentration.,* https://www.mdpi.com/2079-4991/12/1/31
[http://dx.doi.org/10.3390/nano12010031]

[26]   Qayyum, S.; Oves, M.; Khan, A.U. Obliteration of bacterial growth and biofilm through ROS generation by facilely synthesized green silver nanoparticles. *PLoS One,* **2017**, *12*(8): e0181363.
[http://dx.doi.org/10.1371/journal.pone.0181363] [PMID: 28771501]

[27]   Huang, B.; Liu, X.; Li, Z.; Zheng, Y.; Yeung, K.W.K.; Cui, Z.; Liang, Y.; Zhu, S.; Wu, S. Rapid bacteria capturing and killing by AgNPs/N-CD@ZnO hybrids strengthened photo-responsive xerogel for rapid healing of bacteria-infected wounds. *Chem. Engg. J.,* **2021**, *414*, Article 128805.

[28]   Li, P.; Li, J.; Wu, C.; Wu, Q.; Li, J. Synergistic antibacterial effects of β-lactam antibiotic combined with silver nanoparticles. *Nanotechnology,* **2005**, *16*(9), 1912-1917.
[http://dx.doi.org/10.1088/0957-4484/16/9/082]

[29]   Fayaz, A.M.; Balaji, K.; Girilal, M.; Yadav, R.; Kalaichelvan, P.T.; Venketesan, R. Biogenic synthesis of silver nanoparticles and their synergistic effect with antibiotics: a study against gram-positive and gram-negative bacteria. *Nanomedicine,* **2010**, *6*(1), 103-109.

[http://dx.doi.org/10.1016/j.nano.2009.04.006] [PMID: 19447203]

[30]   Deng, H.; McShan, D.; Zhang, Y.; Sinha, S.S.; Arslan, Z.; Ray, P.C.; Yu, H. Mechanistic study of the synergistic antibacterial activity of combined silver nanoparticles and common antibiotics. *Environ. Sci. Technol.,* **2016**, *50*(16), 8840-8848.
[http://dx.doi.org/10.1021/acs.est.6b00998] [PMID: 27390928]

[31]   Jia, Q.; Shan, S.; Jiang, L.; Wang, Y.; Li, D. Synergistic antimicrobial effects of polyaniline combined with silver nanoparticles. *J. Appl. Polym. Sci.,* **2012**, *125*(5), 3560-3566.
[http://dx.doi.org/10.1002/app.36257]

[32]   Lopez-Carrizales, M.; Velasco, K.I.; Castillo, C.; Flores, A.; Magaña, M.; Martinez-Castanon, G.A.; Martinez-Gutierrez, F. *In vitro* synergism of silver nanoparticles with antibiotics as an alternative treatment in multiresistant uropathogens. *Antibiotics (Basel),* **2018**, *7*(2), 50.
[http://dx.doi.org/10.3390/antibiotics7020050] [PMID: 29921822]

[33]   Smekalova, M.; Aragon, V.; Panacek, A.; Prucek, R.; Zboril, R.; Kvitek, L. Enhanced antibacterial effect of antibiotics in combination with silver nanoparticles against animal pathogens. *Vet. J.,* **2016**, *209*, 174-179.
[http://dx.doi.org/10.1016/j.tvjl.2015.10.032] [PMID: 26832810]

[34]   Mc Shan, D.; Zhang, Y.; Deng, H.; Ray, P.C.; Yu, H. Synergistic antibacterial effect of silver nanoparticles combines with ineffective antibiotics on drug resistant Salmonella typhimurium DT104. *J. Environ. Sci. Health Part C Environ. Carcinog. Ecotoxicol. Rev.,* **2015**, *33*(3), 369-384.
[http://dx.doi.org/10.1080/10590501.2015.1055165]

[35]   Devi, L.S.; Joshi, S.R. Antimicrobial and synergistic effects of silver nanoparticles synthesized using soil fungi of high altitudes of eastern himalaya. *Mycobiology,* **2012**, *40*(1), 27-34.
[http://dx.doi.org/10.5941/MYCO.2012.40.1.027] [PMID: 22783131]

[36]   Panacek, A.; Smekalova, M.; Kilianova, M.; Prucek, R.; Bogdanova, K.; Vecerova, R.; Kolar, M.; Havrdova, M.; Plaza, G.A.; Chojniak, J.; Zboril, R.; Kvitek, L. Strong and nonspecific synergistic antibacterial efficacy of antibiotics combined with silver nanoparticles at very low concentrations showing no cytotoxic effect. *Molecules,* **2016**, *21*(1), 26. [MDPI].
[http://dx.doi.org/10.3390/molecules21010026]

[37]   Naqvi, S.Z.; Kiran, U.; Ali, M.I.; Jamal, A.; Hameed, A.; Ahmed, S.; Ali, N. Combined efficacy of biologically synthesized silver nanoparticles and different antibiotics against multidrug-resistant bacteria. *Int. J. Nanomedicine,* **2013**, *8*, 3187-3195.
[http://dx.doi.org/10.2147/IJN.S49284] [PMID: 23986635]

[38]   Mazur, P.; Skiba-Kurek, I.; Mrowiec, P.; Karczewska, E.; Drożdż, R. Synergistic ROS-associated antimicrobial activity of silver nanoparticles and gentamycin against *Staphylococcus epidermidis. Int. J. Nanomedicine,* **2020**, *15*, 3551-3562.
[http://dx.doi.org/10.2147/IJN.S246484] [PMID: 32547013]

[39]   Ghosh, I.N.; Patil, S.D.; Sharma, T.K.; Srivastava, S.K.; Pathania, R.; Navani, N.K. Synergistic action of cinnamaldehyde with silver nanoparticles against spore-forming bacteria: a case for judicious use of silver nanoparticles for antibacterial applications. *Int. J. Nanomedicine,* **2013**, *8*, 4721-4731.
[PMID: 24376352]

[40]   Habash, M.B.; Park, A.J.; Vis, E.C.; Harris, R.J.; Khursigara, C.M. Synergy of silver nanoparticles and aztreonam against Pseudomonas aeruginosa PAO1 biofilms. *Antimicrob. Agents Chemother.,* **2014**, *58*(10), 5818-5830.
[http://dx.doi.org/10.1128/AAC.03170-14] [PMID: 25049240]

[41]   Krychowiak, M.; Kawiak, A.; Narajczyk, M.; Borowik, A.; Królicka, A. Silver nanoparticles combined with naphthoquninones as an effective synergistic strategy against *Staphylococcus aureus. Front. Pharmacol.,* **2018**, *9*, 816.
[http://dx.doi.org/10.3389/fphar.2018.00816] [PMID: 30140226]

[42]   Abo-Shama, U.H.; El-Gendy, H.; Mousa, W.S.; Hamouda, R.A.; Yousuf, W.E.; Hetta, H.F.; Abdeen,

E.E. Synergistic and Antagonistic Effects of Metal Nanoparticles in Combination with Antibiotics Against Some Reference Strains of Pathogenic Microorganisms. *Infect. Drug Resist.,* **2020**, *13*, 351-362.
[http://dx.doi.org/10.2147/IDR.S234425] [PMID: 32104007]

[43] Satapathy, S.; Kumar, S.; Sukhdane, K.S.; Shukla, S.P. Biogenic synthesis and characterization of silver nanoparticles and their effects against bloom-forming algae and synergistic effect with antibiotics against fish pathogenic bacteria. *J. Appl. Phycol.,* **2017**, *29*(4), 1865-1875.
[http://dx.doi.org/10.1007/s10811-017-1091-9]

[44] Lin, P.; Wang, F-Q.; Li, C.T.; Yan, C-F. An Enhancement of Antibacterial Activity and Synergistic Effect of Biosynthesized Silver Nanoparticles by Eurotium cristatum with Various Antibiotics. *Biotechnol. Bioprocess Eng.; BBE,* **2020**, *25*(3), 450-458.
[http://dx.doi.org/10.1007/s12257-019-0506-7]

[45] Vazquez-Muñoz, R.; Meza-Villezcas, A.; Fournier, P.G.J.; Soria-Castro, E.; Juarez-Moreno, K.; Gallego-Hernández, A.L.; Bogdanchikova, N.; Vazquez-Duhalt, R.; Huerta-Saquero, A. Enhancement of antibiotics antimicrobial activity due to the silver nanoparticles impact on the cell membrane. *PLoS One,* **2019**, *14*(11): e0224904.
[http://dx.doi.org/10.1371/journal.pone.0224904] [PMID: 31703098]

[46] Ipe, D.S.; Kumar, P.T.S.; Love, R.M.; Hamlet, S.M. Silver Nanoparticles at Biocompatible Dosage Synergistically Increases Bacterial Susceptibility to Antibiotics. *Front. Microbiol.,* **2020**, *11*, 1074.
[http://dx.doi.org/10.3389/fmicb.2020.01074] [PMID: 32670214]

[47] Asghar, M.A.; Yousuf, R.I.; Shoaib, M.H.; Asghar, M.A.; Ansar, S.; Zehravi, M.; Abdul Rehman, A. Synergistic Nanocomposites of Different Antibiotics Coupled with Green Synthesized Chitosan-Based Silver Nanoparticles: Characterization, Antibacterial, *in vivo* Toxicological and Biodistribution Studies. *Int. J. Nanomedicine,* **2020**, *15*, 7841-7859.
[http://dx.doi.org/10.2147/IJN.S274987] [PMID: 33116504]

[48] Katva, S.; Das, S.; Moti, H.S.; Jyoti, A.; Kaushik, S. Antibacterial Synergy of Silver Nanoparticles with Gentamicin and Chloramphenicol against *Enterococcus faecalis. Pharmacogn. Mag.,* **2018**, *13* Suppl. 4, S828-S833.
[PMID: 29491640]

[49] Naik, M.M.; Prabhu, M.S.; Samant, S.N.; Naik, P.M.; Shilpa, S. Synergistic Action of Silver Nanoparticles Synthesized from Silver Resistant Estuarine Pseudomonas aeruginosa Strain SN5 with Antibiotics against Antibiotic Resistant Bacterial Human Pathogens. *Thalassas,* **2017**, *33*(1), 73-80.
[http://dx.doi.org/10.1007/s41208-017-0023-4]

[50] Tippayawat, P.; Sapa, V.; Srijampa, S.; Boueroy, P.; Chompoosor, A. D-Maltose coated silver nanoparticles and their synergistic effect in combination with ampicillin. *Monatsh. Chem.,* **2017**, *48*(7), 1197-1203.
[http://dx.doi.org/10.1007/s00706-017-2004-y]

[51] Ahmad, S.; Hameed, A.; Khan, K.; Tauseef, I.; Ali, M.; Sultan, F.; Shahzad, M. Evaluation of synergistic effect of nanoparticles with antibiotics against enteric pathogens. *Appl. Nanosci.,* **2019**, *10*(8), 3337-3340.
[http://dx.doi.org/10.1007/s13204-019-01201-3]

[52] Yu, N.; Wang, X.; Qiu, L.; Cai, T.; Jiang, C.; Sun, Y.; Li, Y.; Peng, H.; Xiong, H. Bacteria-triggered hyaluronan/AgNPs/gentamicin nanocarrier for synergistic bacteria disinfection and wound healing application. *Chem. Engg. J.,* **2020**, *380*, Article 122582.

[53] Prema, P.; Thangapandiyan, S.; Immanuel, G. CMC stabilized nano silver synthesis, characterization and its antibacterial and synergistic effect with broad spectrum antibiotics. *Carbohydr. Polym.,* **2017**, *158*, 141-148.
[http://dx.doi.org/10.1016/j.carbpol.2016.11.083] [PMID: 28024537]

[54] Salman, M.; Rizwana, R.; Khan, H.; Munir, I.; Hamayun, M.; Iqbal, A.; Rehman, A.; Amin, K.;

Ahmed, G.; Khan, M.; Khan, A.; Amin, F.U. Synergistic effect of silver nanoparticles and polymyxin B against biofilm produced by *Pseudomonas aeruginosa* isolates of pus samples *in vitro*. *Artif. Cells Nanomed. Biotechnol.,* **2019**, *47*(1), 2465-2472.
[http://dx.doi.org/10.1080/21691401.2019.1626864] [PMID: 31187657]

[55]   Arivarasan, V.K.; Loganathan, K.; Venkatesan, J.; Chaskar, A.C. 'Synergistic-cidal' effect of amoxicillin conjugated silver nanoparticles against Escherichia coli. *Bionanoscience,* **2021**, *11*(2), 506-517.
[http://dx.doi.org/10.1007/s12668-021-00832-7]

[56]   Nakazato, G.; Gonçalves, M.C.; da Silva das Neves, M.; Kobayashi, R.K.T.; Brocchi, M.; Durán, N. Violacein@Biogenic Ag system: synergistic antibacterial activity against Staphylococcus aureus. *Biotechnol. Lett.,* **2019**, *41*(12), 1433-1437.
[http://dx.doi.org/10.1007/s10529-019-02745-8] [PMID: 31650420]

[57]   Sadozai, S.K.; Khan, S.A.; Karim, N.; Becker, D.; Steinbruck, N.; Gier, S.; Baseer, A.; Breinig, F.; Kickelbick, G.; Schneider, M. Ketoconazole-loaded PLGA nanoparticles and their synergism against *Candida albicans* when combined with silver nanoparticles. *J. Drug Deliv. Sci. Technol.,* **2020**, *56*: 101574.
[http://dx.doi.org/10.1016/j.jddst.2020.101574]

[58]   Ullah, S.; Ahmad, A.; Subhan, F.; Jan, A.; Raza, M.; Khan, A.U.; Rahman, A-U.; Khan, U.A.; Tariq, M.; Yuan, Q. Tobramycin mediated silver nanospheres/graphene oxide composite for synergistic therapy of bacterial infection. *J. Photochem. Photobiol. B,* **2018**, *183*, 342-348.
[http://dx.doi.org/10.1016/j.jphotobiol.2018.05.009] [PMID: 29763756]

[59]   Salarian, A.A.; Mollamahale, Y.B.; Hami, Z.; Rad, M.S.R. Cephalexin nanoparticles: Synthesis, cytotoxicity and their synergistic antibacterial study in combination with silver nanoparticles. *Mater. Chem. Phys.,* **2017**, *198*, 125-130.
[http://dx.doi.org/10.1016/j.matchemphys.2017.05.059]

[60]   Anush, K.; Sushanik, K.; Susanna, T.; Ashkhen, H. Antibacterial Effect of Silver and Iron Oxide Nanoparticles in Combination with Antibiotics on E. coli K12. *Bionanoscience,* **2019**, *9*(3), 587-596.
[http://dx.doi.org/10.1007/s12668-019-00640-0]

[61]   Brasil, M.S.L.; Filgueiras, A.L.; Campos, M.B.; Neves, M.S.L.; Eugenio, M.; Sena, L.A.; Anna, C.B.; Silva, V.L.; Diniz, C.G.; Ana, A.C. Synergism in the Antibacterial Action of Ternary Mixtures Involving Silver Nanoparticles, Chitosan and Antibiotics. *J. Braz. Chem. Soc.,* **2018**, *29*, 2026-2033.
[http://dx.doi.org/10.21577/0103-5053.20180077]

[62]   Enan, E.T.; Ashour, A.A.; Basha, S.; Felemban, N.H.; Gad El-Rab, S.M.F. Antimicrobial activity of biosynthesized silver nanoparticles, amoxicillin, and glass-ionomer cement against*Streptococcus mutans*and*Staphylococcus aureus. Nanotechnology,* **2021**, *32*(21): 215101.
[http://dx.doi.org/10.1088/1361-6528/abe577] [PMID: 33657016]

[63]   Huang, W.; Yan, M.; Duan, H.; Bi, Y.; Cheng, X.; Yu, H. Synergistic Antifungal Activity of Green Synthesized Silver Nanoparticles and Epoxiconazole against *Setosphaeria turcica. J. Nanomater.,* **2020**, *2020*: 9535432.
[http://dx.doi.org/10.1155/2020/9535432]

[64]   Thomas, R.; Jishma, P.; Snigdha, S.; Soumya, K.R.; Mathew, J.; Radhakrishnan, E.K. Enhanced antimicrobial efficacy of biosynthesized silver nanoparticle based antibiotic conjugates. *Inorg. Chem. Commun.,* **2020**, *117*: 107978.
[http://dx.doi.org/10.1016/j.inoche.2020.107978]

[65]   Ranpariya, B.; Salunke, G.; Karmakar, S.; Babiya, K.; Sutar, S.; Kadoo, N.; Kumbhakar, P.; Ghosh, S. Antimicrobial synergy of silver-platinum nanohybrids with antibiotics. *Front. Microbiol.,* **2021**, *11*: 610968.
[http://dx.doi.org/10.3389/fmicb.2020.610968] [PMID: 33597929]

[66]   Wan, G.; Ruan, L.; Yin, Y.; Yang, T.; Ge, M.; Cheng, X. Effects of silver nanoparticles in

combination with antibiotics on the resistant bacteria Acinetobacter baumannii. *Int. J. Nanomedicine,* **2016**, *11*, 3789-3800.
[http://dx.doi.org/10.2147/IJN.S104166] [PMID: 27574420]

[67]   Panáček, A.; Smékalová, M.; Večeřová, R.; Bogdanová, K.; Röderová, M.; Kolář, M.; Kilianová, M.; Hradilová, Š.; Froning, J.P.; Havrdová, M.; Prucek, R.; Zbořil, R.; Kvítek, L. Silver nanoparticles strongly enhance and restore bactericidal activity of inactive antibiotics against multiresistant Enterobacteriaceae. *Colloids Surf. B Biointerfaces,* **2016**, *142*, 392-399.
[http://dx.doi.org/10.1016/j.colsurfb.2016.03.007] [PMID: 26970828]

[68]   Jyoti, K.; Baunthiyal, M.; Singh, A. Characterization of silver nanoparticles synthesized using *Urtica dioica* Linn. leaves and their synergistic effects with antibiotics. *J. Radiat. Res. Appl. Sci,* **2016**, *9*(3), 217-227.
[http://dx.doi.org/10.1016/j.jrras.2015.10.002]

[69]   Hassan, K.T.; Ibraheem, I.J.; Hassan, O.M.; Obaid, A.S.; Ali, H.H.; Salih, T.A.; Kadhim, M.S. Facile green synthesis of Ag/AgCl nanoparticles derived from Chara algae extract and evaluating their antibacterial activity and synergistic effect with antibiotics. *J. Environ. Chem. Engg.,* **2021**, *9*, Article 105359.

[70]   Ma, K.; Dong, P.; Liang, M.; Yu, S.; Chen, Y.; Wang, F. Facile Assembly of Multifunctional Antibacterial Nanoplatform Leveraging Synergistic Sensitization between Silver Nanostructure and Vancomycin. *ACS Appl. Mater. Interfaces,* **2020**, *12*(6), 6955-6965.
[http://dx.doi.org/10.1021/acsami.9b22043] [PMID: 31977179]

[71]   Zhou, W.; Jia, Z.; Xiong, P.; Yan, J.; Li, Y.; Li, M.; Cheng, Y.; Zheng, Y. Bioinspired and Biomimetic AgNPs/Gentamicin-Embedded Silk Fibroin Coatings for Robust Antibacterial and Osteogenetic Applications. *ACS Appl. Mater. Interfaces,* **2017**, *9*(31), 25830-25846.
[http://dx.doi.org/10.1021/acsami.7b06757] [PMID: 28731325]

[72]   de Lacerda Coriolano, D.; de Souza, J.B.; Bueno, E.V.; Medeiros, S.M.F.R.D.S.; Cavalcanti, I.D.L.; Cavalcanti, I.M.F.; Cavalcanti, I.M.F. Antibacterial and antibiofilm potential of silver nanoparticles against antibiotic-sensitive and multidrug-resistant Pseudomonas aeruginosa strains. *Braz. J. Microbiol.,* **2021**, *52*(1), 267-278.
[http://dx.doi.org/10.1007/s42770-020-00406-x] [PMID: 33231865]

[73]   Mohamed, M.S.M.; Mostafa, H.M.; Mohamed, S.H.; Abd El-Moez, S.I.; Kamel, Z. Combination of Silver Nanoparticles and Vancomycin to Overcome Antibiotic Resistance in Planktonic/Biofilm Cell from Clinical and Animal Source. *Vet. Microbiol.,* **2020**, *26*(11), 1410-1420.
[http://dx.doi.org/10.1089/mdr.2020.0089] [PMID: 32354252]

[74]   Thomas, R.; Nair, A.P.; Kr, S.; Mathew, J.; Ek, R. Antibacterial activity and synergistic effect of biosynthesized AgNPs with antibiotics against multidrug-resistant biofilm-forming coagulase-negative staphylococci isolated from clinical samples. *Appl. Biochem. Biotechnol.,* **2014**, *173*(2), 449-460.
[http://dx.doi.org/10.1007/s12010-014-0852-z] [PMID: 24699812]

[75]   Rastogi, L.; Kora, A.J.; Sashidhar, R.B. Antibacterial effects of gum kondagogu reduced/stabilized silver nanoparticles in combination with various antibiotics: a mechanistic approach. *Appl. Nanosci.,* **2015**, *5*(5), 535-543.
[http://dx.doi.org/10.1007/s13204-014-0347-9]

[76]   Verma, S.; Abirami, S.; Mahalakshmi, V. Anticancer and antibacterial activity of silver nanoparticles biosynthesized by *Penicillium spp*. and its synergistic effect with antibiotics. *J. Microbiol. Biotechnol. Res.,* **2013**, *3*, 54-71.

[77]   Dhas, S.P.; Mukherjee, A.; Chandrasekharan, N. Synergistic effect of biogenic silver nanocolloid in combination with antibiotics: A potent therapeutic agent. *Int. J. Pharm. Pharm. Sci.,* **2013**, *5*, 292-295.

[78]   Singh, R.; Wagh, P.; Wadhwani, S.; Gaidhani, S.; Kumbhar, A.; Bellare, J.; Chopade, B.A. Synthesis, optimization, and characterization of silver nanoparticles from Acinetobacter calcoaceticus and their enhanced antibacterial activity when combined with antibiotics. *Int. J. Nanomedicine,* **2013**, *8*, 4277-

4290.
[PMID: 24235826]

[79]   Padalia, H.; Moteria, P.; Chanda, S. Green synthesis of silver nanoparticles from marigold flower and its synergistic antimicrobial potential. *Arab. J. Chem.,* **2015**, *8*(5), 732-741.
[http://dx.doi.org/10.1016/j.arabjc.2014.11.015]

[80]   Saratale, G.D.; Saratale, R.G.; Benelli, G.; Kumar, G.; Pugazhendhi, A.; Kim, D-S.; Shin, H-S. Anti-diabetic Potential of Silver Nanoparticles Synthesized with *Argyreia nervosa* Leaf Extract High Synergistic Antibacterial Activity with Standard Antibiotics Against Foodborne Bacteria. *J. Cluster Sci.,* **2017**, *28*(3), 1709-1727.
[http://dx.doi.org/10.1007/s10876-017-1179-z]

[81]   Akram, F.E.; El-Tayeb, T.; Abou-Aisha, K.; El-Azizi, M. A combination of silver nanoparticles and visible blue light enhances the antibacterial efficacy of ineffective antibiotics against methicillin-resistant Staphylococcus aureus (MRSA). *Annal. Clin. Microbiol. Antimicrobial.,* **2016**, *15*, Article 48.

[82]   Singh, P.; Kim, Y.J.; Singh, H.; Mathiyalagan, R.; Wang, C.; Yang, D.C. Biosynthesis of Anisotropic Silver Nanoparticles by *Bhargavaea indica* and Their Synergistic Effect with Antibiotics against Pathogenic Microorganisms. *J. Nanomater.,* **2015**, *2015*: 234741.
[http://dx.doi.org/10.1155/2015/234741]

[83]   Taufeeq, S.M.; Maaroof, M.N.; Al-Ogaidi, I. Synergistic effect of biosynthesized silver nanoparticles with antibiotics against multi-drug resistant bacteria isolated from children with diarrhoea under five years. *Iraqi. J. Sci.,* **2017**, *58*, 41-52.

[84]   Wypij, M.; Świecimska, M.; Czarnecka, J.; Dahm, H.; Rai, M.; Golinska, P. Antimicrobial and cytotoxic activity of silver nanoparticles synthesized from two haloalkaliphilic actinobacterial strains alone and in combination with antibiotics. *J. Appl. Microbiol.,* **2018**, *124*(6), 1411-1424.
[http://dx.doi.org/10.1111/jam.13723] [PMID: 29427473]

SUBJECT INDEX

A

Acid(s) 7, 31, 45, 48, 49, 51, 52, 56, 57, 69, 71, 73, 74, 86, 90, 94, 108, 116
  acrylic 74
  arginylglycylaspartic 48
  ascorbic 7
  carbamoyl-aspartic 31
  carboxylic 56
  chlorogenic 71
  eicosenoic 69
  fatty 31, 51
  folic 90
  Fusidic 116
  glucuronic 52
  hyaluronic 52
  linoleic 69
  mercaptohexanoic 69
  mercaptopropanoic 49
  nucleic 45, 52, 69, 86, 94
  phosphomolybdic 57
  phosphotungstic 57
  phenolic 7
  tannic 108
  thioalkyl 68
  thioctic 69, 73
  thiomalic 69, 73
Acinetobacter baumannii 48, 108, 115, 116
Actinobacteria 25
Aeromonas hydrophila 114
Agents 23, 52, 55, 78, 86
  contrast enhancement 86
  cross- linking 52
  image-enhancing 78
  theranostic 55
Agroecosystems 2, 45
Amino acids 31, 45, 55, 56, 57
  zwitterionic tyrosine 57
Amino-silanized glass surface 109
Amoxicillin 114, 115, 116
AMPs 48, 49
  cationic 49

Amylose 51, 75, 76
  acetylated 51
  encapsulating 76
  helix 76
Anisotropic crystalline structure 25
Antibacterial effect 110, 113
  synergistic 113
Antibiofilm activity 106, 109
Antibiotics 105, 106, 108, 109, 111, 112, 113, 114, 115, 116, 117
  contemporary 105, 117
  neomycin 112
  next-generation 106
Anticancer 55, 86, 89, 98
  chemotherapeutics 86
  drug camptothecin 89
  effect 86, 98
  therapy 55
Anticancer activity 96, 98
  displayed marked 96
Antifungal efficacy 108
Anti-inflammatory activities 52
Antimicrobial 2, 48, 49, 50, 76, 77, 89, 105, 111, 112, 117
  activity 2, 49, 112
  coatings 77
  development programs 117
  peptides 48
  properties 76, 111, 117
  resistance 89
  therapeutics 105
Antitumor effect 50
Apoptosis 46, 87, 88, 89, 91, 92, 96, 97
  inducing 91
  induction of 87, 91, 92
  mitochondria-based 92
Apoptotic cell death 97
Application in gene transfection 94
Applications 7, 8, 54, 86, 92, 95, 96
  anticancer 95, 96
  photonic 8
  sensing 7

www.ingramcontent.com/pod-product-compliance
Lightning Source LLC
Chambersburg PA
CBHW041715210326
41598CB00007B/664